NUTRITIONAL BIOCHEMISTRY

Current Topics
in Nutrition Research

NUTRITIONAL BIOCHEMISTRY

Current Topics in Nutrition Research

Edited by
Chad Cox, PhD

Apple Academic Press Inc. | Apple Academic Press Inc.
3333 Mistwell Crescent | 9 Spinnaker Way
Oakville, ON L6L 0A2 | Waretown, NJ 08758
Canada | USA

©2016 by Apple Academic Press, Inc.

First issued in paperback 2021

Exclusive worldwide distribution by CRC Press, a member of Taylor & Francis Group

No claim to original U.S. Government works

ISBN 13: 978-1-77463-561-2 (pbk)
ISBN 13: 978-1-77188-145-6 (hbk)

Library and Archives Canada Cataloguing in Publication

Nutritional biochemistry : current topics in nutrition research/edited by Chad Cox, PhD.

Includes bibliographical references and index.
ISBN 978-1-77188-145-6 (bound)
1. Nutrition. 2. Metabolism. I. Cox, Chad (Chad L.), author, editor

QP141.N8875 2015   612.3'9   C2015-901402-6

Library of Congress Cataloging-in-Publication Data

Nutritional biochemistry: current topics in nutrition research/editor, Chad Cox.

p. ; cm.
Includes bibliographical references and index.
ISBN 978-1-77188-145-6 (alk. paper)
I. Cox, Chad (Chad L.), editor.
[DNLM: 1. Nutritional Physiological Phenomena. 2. Biochemical Phenomena. 3. Diet Therapy--methods. 4. Nutrition Disorders--etiology. QU 145]

QP141   612.3--dc23   2015006559

Apple Academic Press also publishes its books in a variety of electronic formats. Some content that appears in print may not be available in electronic format. For information about Apple Academic Press products, visit our website at **www.appleacademicpress.com** and the CRC Press website at **www.crc-press.com**

About the Editor

CHAD COX, PhD

Dr. Chad L. Cox is a Lecturer in the Department of Chemistry and the Department of Family and Consumer Sciences at California State University, Sacramento. He also teaches at Sacramento City College, the University of Phoenix, and the University of California, Davis. He holds a PhD in Nutritional Biology, a Bachelor of Science in Exercise Biology, and a Bachelor of Science in Nutrition Science, all from UC Davis. His research interests include the causes of obesity and obesity-related chronic diseases, how exercise training can induce changes in the regulation of gene expression that can lead to improvements in insulin sensitivity and promote energy balance, and the development of pharmacological agents that could help reduce the epidemic of obesity, Type 2 diabetes, and metabolic syndrome.

Contents

Part III: Vitamins and Minerals

Part IV: Macronutrient Composition, Energetics, and Energy Balance

Part V: Cell Function and Metabolism

Acknowledgment and How to Cite

The editor and publisher thank each of the authors who contributed to this book. The chapters in this book were previously published in various places in various formats. To cite the work contained in this book and to view the individual permissions, please refer to the citation at the beginning of each chapter. Each chapter was read individually and carefully selected by the editor; the result is a book that provides a nuanced look at current nutrition research. The chapters included are broken into five sections, which describe the following topics:

- Chapter 1 offers an introductory overview, stressing that gene-nutrient interactions in growth and development and in disease prevention are fundamental to health. It makes the point that nutritional security should be given the same priority as food security.
- Chapter 2 discusses the potential role of nutritional genomics in the current surge of chronic disease in the Middle East and North Africa which has manifested due to the abrupt transition from a traditional to an industrialized diet in these regions. This situation provides an ideal opportunity to apply our current knowledge of nutritional genomics in an effort to help understand and ameliorate an emerging and imminent nutritional disparity.
- A follow-up to their previous pivotal findings in older, overweight subjects (Stanhope et al, May 2009, JCI), in Chapter 3 the Havel group provides additional important evidence regarding the role of sugar consumption in the current epidemic of obesity and chronic disease in young healthy individuals.
- With the diagnosis of metabolic syndrome becoming common in an ever-increasing overweight, aged population, the need for effective interventions is imminent. Dietary restriction and exercise can be effective interventions for those who are motivated; Chapter 4 provides important new information regarding protein requirements in these individuals and how changes in protein intake can affect the efficacy of this type of intervention.
- Chapter 5 is an interesting study comparing the efficacy of oral nutritional supplements with that of nutrition education in patients with Alzheimer's disease. As the diagnosis of dementia becomes more common, it is important to assess the most practical way to deal with the nutritional challenges of this growing population group.

- Vitamin D deficiency has become more prevalent worldwide and it is well-accepted that there is a need for supplementation in susceptible groups; however there is still much debate as to what the appropriate dosage should be. Chapter 6 provides important information regarding the pharmacokinetics of oral vitamin D supplements, and also compares outcomes in both pregnant and non-pregnant women.
- Chapter 7 is an interesting study of the role of vitamin D deficiency in the pathophysiology of childhood obesity and provides evidence that nutritional status, not just body weight, may influence levels of inflammatory mediators and insulin sensitivity in this unfortunately growing population group.
- The success of folic acid fortification in reducing the incidence of neural tube defects is considered one of the success stories of modern nutrition science. Chapter 8 presents an elegant study which provides evidence that increased folic acid intake may also reduce the prevalence of other neuro-degenerative disorders such as autism.
- Chapter 9, authored by the editor of this volume, is important because it shows that, when consumed at 25% of energy requirements, fructose can actually cause the liver to synthesize so much fat after a meal that fat synthesis actually surpasses net whole-body fat oxidation, leading to a temporary state of negative net fat oxidation in body. These data provide additional insights into the mechanisms by which increased sugar consumption leads to the development of obesity.
- High-protein diets are currently a popular fad perhaps in part because of the evidence that this type of diet may lead to reductions in circulating triglyceride levels. In Chapter 10, Oliver and colleagues attempt to determine whether exercise in combination with a high-protein diet has differential effects on triglyceride levels in women with either normal or elevated fasting triglyceride levels.
- Tea is one of the most commonly consumed beverages globally and there have been many studies examining the potential health benefits of various types of teas. In Chapter 11, Rajavelu and colleagues provide data suggesting a potential biochemical mechanism by which compounds found in black tea and coffee may modulate the methylation of DNA.
- Supplement use is at an all-time high in the United States and finding peer-reviewed data validating the efficacy of many of these products can be very difficult. Chapter 12 examines the efficacy of several commercially available supplement products and measures their ability to modulate the mitochondrial content of human rhabdomyosarcoma cells.

List of Contributors

Abdullah Al Mahmud
International Center for Diarrhoeal Disease Research, Bangladesh (icddr,b), Dhaka, Bangladesh

Nigel Arden
MRC Lifecourse Epidemiology Unit, University of Southampton, Southampton, United Kingdom and NIHR Musculoskeletal BRU, Botnar Research Centre, Oxford, United Kingdom

Abdullah H. Baqui
Department of International Health, The Johns Hopkins Bloomberg School of Public Health, Baltimore, MD 21205, USA and International Center for Diarrhoeal Disease Research, Bangladesh (icddr,b), Dhaka, Bangladesh

Miguel A. Barberena
Department of Biochemistry and Molecular Biology, University of New Mexico Health Sciences Center, Albuquerque, NM, 87131, USA

L. Berglund
Department of Internal Medicine, UCD School of Medicine, Sacramento, CA, USA

Diane J. Berry
Centre for Paediatric Epidemiology and Biostatistics and MRC Centre of Epidemiology for Child Health, UCL Institute of Child Health, London, United Kingdom

Paulo H. F. Bertolucci
Department Neurology and Neurosurgery, Behaviour Neurology Section, Universidade Federal de São Paulo/UNIFESP-EPM, São Paulo, Brazil

Marco Bisoffi
Department of Biochemistry and Molecular Biology, University of New Mexico Health Sciences Center, Albuquerque, NM, 87131, USA

Robert E. Black
Department of International Health, The Johns Hopkins Bloomberg School of Public Health, Baltimore, MD 21205, USA

Peter G. Bourne
Green Templeton College, University of Oxford, Oxford OX2 6HG, UK

César Q. Brant
Department Neurology and Neurosurgery, Behaviour Neurology Section, Universidade Federal de São Paulo/UNIFESP-EPM, São Paulo, Brazil

Andrew A. Bremer
Department of Pediatrics, University of California, Davis and Department of Pediatrics, School of Medicine, Vanderbilt University, Nashville, Tennessee 37204

Liisa Byberg
Department of Surgical Sciences, Uppsala University, Uppsala, Sweden

Mike Byrd
Exercise and Sports Nutrition Laboratory, Texas A & M University, College Station, TX, USA

Harry Campbell
Centre for Population Health Sciences, University of Edinburgh, Edinburgh, United Kingdom

Claire Canon
Exercise and Sports Nutrition Laboratory, Texas A & M University, College Station, TX, USA

Robert Chapier
Clermont University, Blaise Pascal University, Laboratory of Metabolic Adaptations to Exercise in Physiological and Pathological conditions (AME2P, EA3533), BP 10448, F-63000, Clermont-Ferrand, France

Guoxia Chen
Department of Molecular Biosciences, University of California, Davis

Carole A. Conn
Department of IFCE: Nutrition, University of New Mexico, Albuquerque, NM, 87131, USA

Cyrus Cooper
MRC Lifecourse Epidemiology Unit, University of Southampton, Southampton, United Kingdom

Jason D. Cooper
Juvenile Diabetes Research Foundation/Wellcome Trust Diabetes and Inflammation Laboratory, Department of Medical Genetics, Cambridge Institute for Medical Research, University of Cambridge, Cambridge, United Kingdom

Daniel Courteix
Clermont University, Blaise Pascal University, Laboratory of Metabolic Adaptations to Exercise in Physiological and Pathological conditions (AME2P, EA3533), BP 10448, F-63000, Clermont-Ferrand, France and School of Exercise Science, Australian Catholic University, Locked Bag 4115 Fitzroy MDC VIC 3165, Australia

C. L. Cox
Department of Nutrition, University of California, Davis, Davis, CA, USA

Rosimeire V. da Silva
Department Neurology and Neurosurgery, Behaviour Neurology Section, Universidade Federal de São Paulo/UNIFESP-EPM, São Paulo, Brazil

Zari Dastani
Department of Epidemiology, Biostatistics and Occupational Health, Lady Davis Institute, Jewish General Hospital, McGill University, Montreal, Quebec, Canada

Eric Doré
Clermont University, Blaise Pascal University, Laboratory of Metabolic Adaptations to Exercise in Physiological and Pathological conditions (AME2P, EA3533), BP 10448, F-63000, Clermont-Ferrand, France

Malcolm Dunlop
Colon Cancer Genetics Group and Academic Coloproctology, Institute of Genetics and Molecular Medicine, University of Edinburgh, United Kingdom and MRC Human Genetics Unit Western General Hospital Edinburgh, United Kingdom

Josée Dupuis
Department of Biostatistics, Boston University School of Public Health, Boston, Massachusetts, United States of America and National Heart, Lung, and Blood Institute's Framingham Heart Study, Framingham, Massachusetts, United States of America

Frédéric Dutheil
Clermont University, Blaise Pascal University, Laboratory of Metabolic Adaptations to Exercise in Physiological and Pathological conditions (AME2P, EA3533), BP 10448, F-63000, Clermont-Ferrand, France, Sport Medicine and Functional Explorations, University Hospital (CHU) G. Montpied, F-63000, Clermont-Ferrand, France, and Occupational Medicine, Faculty of Medicine, F-63000, Clermont-Ferrand, France

Abdul-Karim M. El-Hage-Sleiman
Department of Biochemistry and Molecular Genetics, American University of Beirut, Beirut 1107 2020, Lebanon

Ole Faergeman
Department of Internal Medicine and Cardiology, Aarhus Sygehus University Hospital Tage Hansens Gade 2, 8000 Aarhus C, Denmark

Akl C. Fahed
Department of Genetics, Harvard Medical School, 77 Avenue Louis Pasteur, Boston, MA 02115, USA

Theresa I. Farhat
Department of Biochemistry and Molecular Genetics, American University of Beirut, Beirut 1107 2020, Lebanon

Luigi Ferrucci
Clinical Research Branch, Harbor Hospital, Baltimore, Maryland, United States of America

Tak Hou Fong
Department of Molecular Biosciences, University of California, Davis

Timothy M. Frayling
Genetics of Complex Traits, Peninsula College of Medicine and Dentistry, University of Exeter, Exeter, United Kingdom

Randi Garcia-Smith
Department of Biochemistry and Molecular Biology, University of New Mexico Health Sciences Center, Albuquerque, NM, 87131, USA

J. L. Graham
Department of Nutrition, University of California, Davis, Davis, CA, USA and Department of Molecular Biosciences, School of Veterinary Medicine, University of California, Davis, Davis, CA, USA

S. C. Griffen
Department of Internal Medicine, UCD School of Medicine, Sacramento, CA, USA

Anna-Liisa Hartikainen
Department of Obstetrics and Gynaecology and Public Health and General Practice, University of Oulu, Oulu, Finland

B. Hatcher
Department of Nutrition, University of California, Davis, Davis, CA, USA

Peter J. Havel
Department of Molecular Biosciences and School of Veterinary Medicine, and Department of Nutrition, University of California, Davis

Karl-Heinz Herzig
Institute of Health Sciences and Biocenter Oulu, University of Oulu, Oulu, Finland and Institute of Biomedicine, University of Oulu, Oulu, Finland and Department of Psychiatry, Kuopio University Hospital, Kuopio, Finland

Aroon D. Hingorani
Genetic Epidemiology Group, Department of Epidemiology and Public Health, Division of Population Health, University College London, London, United Kingdom and Division of Medicine, Centre for Clinical Pharmacology, University College London, London, United Kingdom

Linda T. Hiraki
Program in Molecular and Genetic Epidemiology, Harvard School of Public Health, Boston, Massachusetts, United States of America

Denise K. Houston
Department of Internal Medicine, Section on Gerontology and Geriatric Medicine, Wake Forest School of Medicine, Winston Salem, North Carolina, United States of America

Elina Hyppönen
Centre for Paediatric Epidemiology and Biostatistics and MRC Centre of Epidemiology for Child Health, UCL Institute of Child Health, London, United Kingdom

Erik Ingelsson
Department of Medical Epidemiology and Biostatistics, Karolinska Institutet, Stockholm, Sweden

Yasuki Ito
Denka Seiken Co., Ltd. Tokyo103-0025, Japan

Rakesh Jaiswal
Chemistry, Jacobs University Bremen, Campus Ring 1, 28759 Bremen, Germany

Karen Jameson
MRC Lifecourse Epidemiology Unit, University of Southampton, Southampton, United Kingdom

Marjo-Riitta Järvelin
Institute of Health Sciences and Biocenter Oulu, University of Oulu, Oulu, Finland, Department of Biostatistics and Epidemiology, School of Public Health, MRC-HPA Centre for Environment and Health, Imperial College, Faculty of Medicine, London, United Kingdom and Department of Children, Young People and Families, National Institute for Health and Welfare, Oulu, Finland

Albert Jeltsch
Biochemistry, Jacobs University Bremen, Campus Ring 1, 28759 Bremen, Germany

Antti Jula
National Institute for Health and Welfare, Helsinki, Finland

Yara Juliano
Department of Nutrition, Universidade de Santo Amaro/UNISA, São Paulo, Brazil

Peter Jung
Exercise and Sports Nutrition Laboratory, Texas A & M University, College Station, TX, USA

Marika Kaakinen
Institute of Health Sciences and Biocenter Oulu, University of Oulu, Oulu, Finland

Nancy L. Keim
School of Veterinary Medicine, Department of Nutrition, and School of Medicine, University of California, Davis; United States Department of Agriculture

Chad Kerksick
Applied Biochemistry and Molecular Physiology Laboratory, University of Oklahoma, Norman, OK, USA

Deepesh Khanna
Exercise and Sports Nutrition Laboratory, Texas A & M University, College Station, TX, USA

Marcus E. Kleber
LURIC Study non-profit LLC, Freiburg, Germany and Mannheim Institute of Public Health, Social and Preventive Medicine, Mannheim Medical Faculty, University of Heidelberg, Mannheim, Germany

Peter Kraft
Program in Molecular and Genetic Epidemiology, Harvard School of Public Health, Boston, Massachusetts, United States of America

Richard Kreider
Exercise and Sports Nutrition Laboratory, Texas A & M University, College Station, TX, USA

Julie Y. Kresta
Exercise and Sports Nutrition Laboratory, Texas A & M University, College Station, TX, USA

Stephen B. Kritchevsky
Department of Internal Medicine, Section on Gerontology and Geriatric Medicine, Wake Forest School of Medicine, Winston Salem, North Carolina, United States of America

Majid Koozehchian
Exercise and Sports Nutrition Laboratory, Texas A & M University, College Station, TX, USA

Nikolai Kuhnert
Chemistry, Jacobs University Bremen, Campus Ring 1, 28759 Bremen, Germany

Gérard Lac
Clermont University, Blaise Pascal University, Laboratory of Metabolic Adaptations to Exercise in Physiological and Pathological conditions (AME2P, EA3533), BP 10448, F-63000, Clermont-Ferrand, France

Vivien Lee
Department of Molecular Biosciences, University of California, Davis

Terho Lehtimäki
Department of Clinical Chemistry, Fimlab Laboratories, Tampere University Hospital and University of Tampere, Tampere, Finland

Bruno Lesourd
Clermont University, Blaise Pascal University, Laboratory of Metabolic Adaptations to Exercise in Physiological and Pathological conditions (AME2P, EA3533), BP 10448, F-63000, Clermont-Ferrand, France and Geriatrics Departments, Faculty of Medicine, F-63000, Clermont-Ferrand, France

Rui Li
Departments of Medicine, Human Genetics, Epidemiology and Biostatistics, Lady Davis Institute, Jewish General Hospital, McGill University, Montreal, Quebec, Canada

Lars Lind
Department of Medical Sciences, Uppsala University, Uppsala, Sweden

Brittanie Lockard
Exercise and Sports Nutrition Laboratory, Texas A & M University, College Station, TX, USA

Kurt K. Lohman
Department of Biostatistical Sciences, Division of Public Health Sciences, Wake Forest School of Medicine, Winston Salem, North Carolina, United States of America

Ruth J. F. Loos
MRC Epidemiology Unit, Institute of Metabolic Science, Addenbrooke's Hospital, Cambridge, United Kingdom

Mattias Lorentzon
Center for Bone and Arthritis Research, Department of Internal Medicine, Institute of Medicine, University of Gothenburg, Gothenburg, Sweden

Chen Lu
Department of Biostatistics, Boston University School of Public Health, Boston, Massachusetts, United States of America

Michelle Mardock
Exercise and Sports Nutrition Laboratory, Texas A & M University, College Station, TX, USA

Winfried März
Synlab Academy, Mannheim, Germany and Mannheim Institute of Public Health, Social and Preventive Medicine, Mannheim Medical Faculty, University of Heidelberg, Mannheim, Germany

Mark I. McCarthy
Oxford Centre for Diabetes, Endocrinology and Metabolism, University of Oxford, Churchill Hospital, Headington, Oxford, United Kingdom, Wellcome Trust Centre for Human Genetics, University of Oxford, Oxford, United Kingdom and Oxford NIHR Biomedical Research Centre, Churchill Hospital, Headington, Oxford, United Kingdom

J. P. McGahan
Department of Radiology, UCD Medical Center, Sacramento, CA, USA

Valentina Medici
Division of Gastroenterology and Hepatology, University of California, Davis

Håkan Melhus
Department of Medical Sciences, Uppsala University, Uppsala, Sweden

Roseanne I. Menorca
Department of Molecular Biosciences, University of California, Davis

Karl Michaëlsson
Department of Surgical Sciences, Uppsala University, Uppsala, Sweden

Braxton D. Mitchell
University of Maryland School of Medicine, Division of Endocrinology, Baltimore, Maryland, United States of America

Katsuyuki Nakajima
Lipid Metabolism Laboratory, Jean Mayer Human Nutrition Research Center on Aging, Tufts University and Tufts University School of Medicine, Boston, Massachusetts 02111 and Diagnostic Division Otsuka Pharmaceutical Co., Ltd., Tokyo 101-8535, Japan

Takamitsu Nakano
Diagnostic Division, Otsuka Pharmaceutical Co., Ltd., Tokyo 101-8535, Japan

Yasmin Neggers
Department of Human Nutrition, University of Alabama, Box 870311, 504 University Blvd, Tuscaloosa, AL 35487, USA

Georges M. Nemer
Department of Biochemistry and Molecular Genetics, American University of Beirut, Beirut 1107 2020, Lebanon

Neil F. Novo
Department of Nutrition, Universidade de Santo Amaro/UNISA, São Paulo, Brazil

Claes Ohlsson
Center for Bone and Arthritis Research, Department of Internal Medicine, Institute of Medicine, University of Gothenburg, Gothenburg, Sweden

Ivan H. Okamoto
Department Neurology and Neurosurgery, Behaviour Neurology Section, Universidade Federal de São Paulo/UNIFESP-EPM, São Paulo, Brazil

Jonathan M. Oliver
Exercise and Sports Nutrition Laboratory, Texas A & M University, College Station, TX, USA

Willem H. Ouwehand
Department of Haematology, University of Cambridge, United Kingdom, Wellcome Trust Sanger Institute, Hinxton, Cambridge, United Kingdom, and NHS Blood and Transplant, Cambridge, United Kingdom

Stefan Pilz
Department of Internal Medicine, Division of Endocrinology and Metabolism, Medical University of Graz, Austria and Department of Epidemiology and Biostatistics, EMGO Institute for Health and Care Research, VU University Medical Centre, Amsterdam, The Netherlands

Glaucia A. K. Pivi
Department Neurology and Neurosurgery, Behaviour Neurology Section, Universidade Federal de São Paulo/UNIFESP-EPM, São Paulo, Brazil

Chris Power
Centre for Paediatric Epidemiology and Biostatistics and MRC Centre of Epidemiology for Child Health, UCL Institute of Child Health, London, United Kingdom

Anneli Pouta
Department of Public Health Science and General Practice, University of Oulu, Oulu, Finland

Olli Raitakari
Research Centre of Applied and Preventive Cardiovascular Medicine, University of Turku and Department of Clinical Physiology and Nuclear Medicine, University of Turku and Turku University Hospital, Turku, Finland

Arumugam Rajavelu
Biochemistry, Jacobs University Bremen, Campus Ring 1, 28759 Bremen, Germany

Rubhana Raqib
International Center for Diarrhoeal Disease Research, Bangladesh (icddr,b), Dhaka, Bangladesh

Chris Rasmussen
Exercise and Sports Nutrition Laboratory, Texas A & M University, College Station, TX, USA

J. Brent Richards
Departments of Medicine, Human Genetics, Epidemiology and Biostatistics, Lady Davis Institute, Jewish General Hospital, McGill University, Montreal, Quebec, Canada and Department of Twin Research and Genetic Epidemiology, King's College London, London, United Kingdom

Laurence Roszyk
Biochemistry, University Hospital (CHU) G. Montpied, F-63000, Clermont-Ferrand, France

Daniel E. Roth
Department of International Health, The Johns Hopkins Bloomberg School of Public Health, Baltimore, MD 21205, USA. Current address: Department of Paediatrics, Hospital for Sick Children and University of Toronto, Toronto, ON, Canada

Veikko Salomaa
National Institute for Health and Welfare, Helsinki, Finland

Vincent Sapin
Biochemistry, University Hospital (CHU) G. Montpied, F-63000, Clermont-Ferrand, France

J. M. Schwarz
Department of Basic Sciences, College of Osteopathic Medicine, Touro University, Vallejo, CA, USA

Sunday Simbo
Exercise and Sports Nutrition Laboratory, Texas A & M University, College Station, TX, USA

Artemis P. Simopoulos
The Center for Genetics, Nutrition and Health, Washington, DC 20009, USA

Timothy D. Spector
Department of Twin Research and Genetic Epidemiology, King's College London, London, United Kingdom

Kimber L. Stanhope
Department of Molecular Biosciences and School of Veterinary Medicine, and Department of Nutrition, University of California, Davis

Elizabeth A. Streeten
University of Maryland School of Medicine, Division of Endocrinology, Baltimore, Maryland, United States of America

Evropi Theodoratou
Centre for Population Health Sciences, University of Edinburgh, Edinburgh, United Kingdom

Emmi Tikkanen
Institute for Molecular Medicine Finland FIMM, University of Helsinki, Helsinki, Finland and National Institute for Health and Welfare, Helsinki, Finland

Kristina Trujillo
Department of Biochemistry and Molecular Biology, University of New Mexico Health Sciences Center, Albuquerque, NM, 87131, USA

Zumrad Tulyasheva
MoLife program, Jacobs University Bremen, Campus Ring 1, 28759 Bremen, Germany

Liesbeth Vandenput
Center for Bone and Arthritis Research, Department of Internal Medicine, Institute of Medicine, University of Gothenburg, Gothenburg, Sweden

Roger A. Vaughan
Department of Health, Exercise and Sports Science, University of New Mexico, University Blvd, Albuquerque, NM, 87131, USA, Department of Biochemistry and Molecular Biology, University of New Mexico Health Sciences Center, Albuquerque, NM, 87131, USA, and Department of IFCE: Nutrition, University of New Mexico, Albuquerque, NM, 87131, USA

Jorma Viikari
Department of Medicine, University of Turku and Turku University Hospital, Turku, Finland

Karani S. Vimaleswaran
Centre for Paediatric Epidemiology and Biostatistics and MRC Centre of Epidemiology for Child Health, UCL Institute of Child Health, London, United Kingdom

Thomas J. Wang
Cardiology Division, Massachusetts General Hospital, Boston, Massachusetts, United States of America

Nicholas J. Wareham
MRC Epidemiology Unit, Institute of Metabolic Science, Addenbrooke's Hospital, Cambridge, United Kingdom

John C. Whittaker
Department of Epidemiology and Population Health, London School of Hygiene and Tropical Medicine, London, United Kingdom and Quantitative Sciences, GlaxoSmithKline, Stevenage, United Kingdom

Andrew R. Wood
Genetics of Complex Traits, Peninsula College of Medicine and Dentistry, University of Exeter, Exeter, United Kingdom

Laura M. Yerges-Armstrong
University of Maryland School of Medicine, Division of Endocrinology, Baltimore, Maryland, United States of America

Lina Zgaga
Centre for Population Health Sciences, University of Edinburgh, Edinburgh, United Kingdom and Andrija Stampar School of Public Health, Medical School University of Zagreb, Zagreb, Croatia

Introduction

Nutrition is becoming ever more central to our understanding of virtually all metabolic processes. Its biological basis offers insight into the mechanisms by which diet influences human health and disease. Nutritional biochemistry broadens and deepens our understanding of many aspects of human biology including immunity, development, and aging. Research in this complex field must integrate information from a myriad of fields including cellular and molecular biology, molecular genetics, physiology, epidemiology, and clinical medicine.

Nutritional biochemistry is a vital field of study. According to the US Department of Health and Human Services, poor nutrition (along with inactivity) contributes to approximately half a million deaths each year. Poor nutrition is also a leading cause of disability and loss of independence. Healthier dietary practices, according to the US Department of Agricuture, could prevent at least $71 billion per year in medical costs, lost productivity, and premature deaths in the United States alone. At the international level, the World Health Organization focuses on nutrition as one of the most significant factors influencing human health; undernutrition contributes to about a third of all child deaths around the world, and growing rates of overweight and obesity worldwide are associated with the increasing prevalence of chronic diseases such as cancer, cardiovascular disease, and diabetes.

The research gathered in this compendium focuses on a few select elements of nutritional biochemistry: vitamins and minerals, macronutrients and energy, cell function and metabolism, as well as clinical nutrition. Each chapter contributes to our larger understanding of this complex field. Together, these articles offer a springboard for ongoing research in a field that is essential for more informed nutritional policies and programs at the individual, clinical, educational, and governmental levels.

Chad Cox, PhD

Chapter 1, by Simopoulos and colleagues, discusses the *Bellagio Report on Healthy Agriculture, Healthy Nutrition, Healthy People*: the result of the meeting held at the Rockefeller Foundation Bellagio Center in Lake Como, Italy, 29 October–2 November 2012. The meeting was science-based but policy-oriented. The role and amount of healthy and unhealthy fats, with attention to the relative content of omega-3 and omega-6 fatty acids, sugar, and particularly fructose in foods that may underlie the epidemics of non-communicable diseases (NCD's) worldwide were extensively discussed. The report concludes that sugar consumption, especially in the form of high energy fructose in soft drinks, poses a major and insidious health threat, especially in children, and most diets, although with regional differences, are deficient in omega-3 fatty acids and too high in omega-6 fatty acids. Gene-nutrient interactions in growth and development and in disease prevention are fundamental to health, therefore regional Centers on Genetics, Nutrition and Fitness for Health should be established worldwide. Heads of state and government must elevate, as a matter of urgency, Nutrition as a national priority, that access to a healthy diet should be considered a human right and that the lead responsibility for Nutrition should be placed in Ministries of Health rather than agriculture so that the health requirements drive agricultural priorities, not vice versa. Nutritional security should be given the same priority as food security.

The Middle East and North Africa (MENA) region suffers a drastic change from a traditional diet to an industrialized diet. This has led to an unparalleled increase in the prevalence of chronic diseases. Chapter 2, by Fahed and colleagues, discusses the role of nutritional genomics, or the dietary signature, in these dietary and disease changes in the MENA. The diet-genetics-disease relation is discussed in detail. Selected disease categories in the MENA are discussed starting with a review of their epidemiology in the different MENA countries, followed by an examination of the known genetic factors that have been reported in the disease discussed, whether inside or outside the MENA. Several diet-genetics-disease relationships in the MENA may be contributing to the increased prevalence of civilization disorders of metabolism and micronutrient deficiencies. Future research in the field of nutritional genomics in the MENA is needed to better define these relationships.

The American Heart Association Nutrition Committee recommends women and men consume no more than 100 and 150 kcal of added sugar per day, respectively, whereas the Dietary Guidelines for Americans, 2010, suggests a maximal added sugar intake of 25% or less of total energy. To address this discrepancy, the authors of Chapter 3, Stanhope and colleagues, compared the effects of consuming glucose, fructose, or high-fructose corn syrup (HFCS) at 25% of energy requirements (E) on risk factors for cardiovascular disease. Forty-eight adults (aged 18–40 yr; body mass index 18–35 kg/m^2) resided at the Clinical Research Center for 3.5 d of baseline testing while consuming energy-balanced diets containing 55% E complex carbohydrate. For 12 outpatient days, they consumed usual ad libitum diets along with three servings per day of glucose, fructose, or HFCS-sweetened beverages (n = 16/group), which provided 25% E requirements. Subjects then consumed energy-balanced diets containing 25% E sugar-sweetened beverages/30% E complex carbohydrate during 3.5 d of inpatient intervention testing. Twenty-four-hour triglyceride area under the curve, fasting plasma low-density lipoprotein (LDL), and apolipoprotein B (apoB) concentrations were measured. Twenty-four-hour triglyceride area under the curve was increased compared with baseline during consumption of fructose (+4.7 ± 1.2 mmol/liter × 24 h, P = 0.0032) and HFCS (+1.8 ± 1.4 mmol/liter × 24 h, P = 0.035) but not glucose (−1.9 ± 0.9 mmol/liter × 24 h, P = 0.14). Fasting LDL and apoB concentrations were increased during consumption of fructose (LDL: +0.29 ± 0.082 mmol/liter, P = 0.0023; apoB: +0.093 ± 0.022 g/liter, P = 0.0005) and HFCS (LDL: +0.42 ± 0.11 mmol/liter, P < 0.0001; apoB: +0.12 ± 0.031 g/liter, P < 0.0001) but not glucose (LDL: +0.012 ± 0.071 mmol/liter, P = 0.86; apoB: +0.0097 ± 0.019 g/liter, P = 0.90). Consumption of HFCS-sweetened beverages for 2 wk at 25% E increased risk factors for cardiovascular disease comparably with fructose and more than glucose in young adults.

The recommended dietary allowance (RDA) for protein intake has been set at 1.0–1.3 g/kg/day for seniors. To date, no consensus exists on the lower threshold intake (LTI=RDA/1.3) for the protein intake (PI) needed in senior patients ongoing both combined caloric restriction and physical activity treatment for metabolic syndrome. Considering that age, caloric restriction and exercise are three increasing factors of protein need,

Chapter 4, by Dutheil and colleagues, was dedicated to determining the minimal PI in this situation, through the determination of albuminemia that is the blood marker of protein homeostasis. Twenty eight subjects (19 M, 9 F, 61.8 ± 6.5 years, BMI 33.4 ± 4.1 kg/m^2) with metabolic syndrome completed a three-week residential programme (Day 0 to Day 21) controlled for nutrition (energy balance of -500 kcal/day) and physical activity (3.5 hours/day). Patients were randomly assigned in two groups: Normal-PI (NPI: 1.0 g/kg/day) and High-PI (HPI: 1.2 g/kg/day). Then, patients returned home and were followed for six months. Albuminemia was measured at D0, D21, D90 and D180. At baseline, PI was spontaneously 1.0 g/kg/day for both groups. Albuminemia was 40.6 g/l for NPI and 40.8 g/l for HPI. A marginal protein under-nutrition appeared in NPI with a decreased albuminemia at D90 below 35 g/l (34.3 versus 41.5 g/l for HPI, $p < 0.05$), whereas albuminemia remained stable in HPI. During the treatment based on restricted diet and exercise in senior people with metabolic syndrome, the lower threshold intake for protein must be set at 1.2 g/kg/day to maintain blood protein homeostasis.

Weight loss in patients with Alzheimer's disease (AD) is a common clinical manifestation that may have clinical significance. Chapter 5, by Pivi and colleagues, aimed to evaluate if there is a difference between nutrition education and oral nutritional supplementation on nutritional status in patients with AD. A randomized, prospective 6-month study which enrolled 90 subjects with probable AD aged 65 years or older divided into 3 groups: Control Group (CG) [n = 27], Education Group (EG) [n = 25], which participated in an education program and Supplementation Group (SG) [n = 26], which received two daily servings of oral nutritional supplementation. Subjects were assessed for anthropometric data (weight, height, BMI, TSF, AC and AMC), biochemical data (total protein, albumin, and total lymphocyte count), CDR (Clinical Dementia Rating), MMSE (Mini-mental state examination), as well as dependence during meals. The SG showed a significant improvement in the following anthropometric measurements: weight (H calc = 22.12, p =< 0.001), BMI (H calc = 22.12, p =< 0.001), AC (H calc = 12.99, p =< 0.002), and AMC (H calc = 8.67, p =< 0.013) compared to the CG and EG. BMI of the EG was significantly greater compared to the CG. There were significant changes in total protein (H calc = 6.17, p =< 0.046), and total lymphocyte count in the

SG compared to the other groups (H cal = 7.94, p = 0.019). The study concluded that oral nutritional supplementation is more effective compared to nutrition education in improving nutritional status.

Improvements in antenatal vitamin D status may have maternal-infant health benefits. To inform the design of prenatal vitamin D3 trials, the authors of Chapter 6, Roth and colleagues, conducted a pharmacokinetic study of single-dose vitamin D3 supplementation in women of reproductive age. A single oral vitamin D3 dose (70,000 IU) was administered to 34 non-pregnant and 27 pregnant women (27 to 30 weeks gestation) enrolled in Dhaka, Bangladesh (23°N). The primary pharmacokinetic outcome measure was the change in serum 25-hydroxyvitamin D concentration over time, estimated using model-independent pharmacokinetic parameters. Baseline mean serum 25-hydroxyvitamin D concentration was 54 nmol/L (95% CI 47, 62) in non-pregnant participants and 39 nmol/L (95% CI 34, 45) in pregnant women. Mean peak rise in serum 25-hydroxyvitamin D concentration above baseline was similar in non-pregnant and pregnant women (28 nmol/L and 32 nmol/L, respectively). However, the rate of rise was slightly slower in pregnant women (i.e., lower 25-hydroxyvitamin D on day 2 and higher 25-hydroxyvitamin D on day 21 versus non-pregnant participants). Overall, average 25-hydroxyvitamin D concentration was 19 nmol/L above baseline during the first month. Supplementation did not induce hypercalcemia, and there were no supplement-related adverse events. The response to a single 70,000 IU dose of vitamin D3 was similar in pregnant and non-pregnant women in Dhaka and consistent with previous studies in non-pregnant adults. These preliminary data support the further investigation of antenatal vitamin D3 regimens involving doses of ≤70,000 IU in regions where maternal-infant vitamin D deficiency is common.

Obesity is associated with vitamin D deficiency, and both are areas of active public health concern. In Chapter 7, Vimaleswaran and colleagues explored the causality and direction of the relationship between body mass index (BMI) and 25-hydroxyvitamin D [25(OH)D] using genetic markers as instrumental variables (IVs) in bi-directional Mendelian randomization (MR) analysis. The authors used information from 21 adult cohorts (up to 42,024 participants) with 12 BMI-related SNPs (combined in an allelic score) to produce an instrument for BMI and four SNPs associated with 25(OH)D (combined in two allelic scores, separately for genes encoding its synthesis or metabolism) as an instrument for vitamin D. Regression estimates for the IVs (allele scores) were generated within-study and pooled by meta-analysis to generate summary effects. Associations between vitamin D scores and BMI were confirmed in the Genetic Investigation of Anthropometric Traits (GIANT) consortium (n = 123,864). Each 1 kg/m² higher BMI was associated with 1.15% lower 25(OH)D (p = 6.52×10^{-27}). The BMI allele score was associated both with BMI (p = 6.30×10^{-62}) and 25(OH)D (−0.06% [95% CI −0.10 to −0.02], p = 0.004) in the cohorts that underwent meta-analysis. The two vitamin D allele scores were strongly associated with 25(OH)D (p$\leq$8.07$\times 10^{-57}$ for both scores) but not with BMI (synthesis score, p = 0.88; metabolism score, p = 0.08) in the meta-analysis. A 10% higher genetically instrumented BMI was associated with 4.2% lower 25(OH)D concentrations (IV ratio: −4.2 [95% CI −7.1 to −1.3], p = 0.005). No association was seen for genetically instrumented 25(OH)D with BMI, a finding that was confirmed using data from the GIANT consortium (p$\geq$0.57 for both vitamin D scores). On the basis of a bi-directional genetic approach that limits confounding, this study suggests that a higher BMI leads to lower 25(OH)D, while any effects of lower 25(OH)D increasing BMI are likely to be small. Population level interventions to reduce BMI are expected to decrease the prevalence of vitamin D deficiency.

Many factors such as genetic, psychological, environmental and behavioral characteristics influence development of human disease, and there is a great deal of scientific evidence supporting a close relationship between nutrition and disease. Though typical Western diets are not overtly deficient in essential nutrients, intake of a few micro nutrients, such as folic acid, have been reported to be sub-optimal, particularly in women of childbearing age. The role of folic acid in the prevention of macrocytic anemia and neural tube defects is well established. However, the relationship between folic acid and risk of autistic spectrum disorder (ASD) is still evolving. In Chapter 8, authored by Yasmin Neggars, the association between maternal folic acid intake, including folic acid supplementaion and risk of ASD in the offspring, will be evaluated. It is of interest that though there is some evidence that folate intake during pregnancy decreases the risk of ASD, several investigators have speculated that a high maternal folate intake due to folic acid fortification of foods may be linked to increased prevalence of ASD.

The results of short-term studies in humans suggest that, compared with glucose, acute consumption of fructose leads to increased postprandial energy expenditure and carbohydrate oxidation and decreased postprandial fat oxidation. The objective of Chapter 9, by Cox and colleagues, was to determine the potential effects of increased fructose consumption compared with isocaloric glucose consumption on substrate utilization and energy expenditure following sustained consumption and under energy-balanced conditions. As part of a parallel arm study, overweight/obese male and female subjects, 40–72 years, consumed glucose- or fructose-sweetened beverages providing 25% of energy requirements for 10 weeks. Energy expenditure and substrate utilization were assessed using indirect calorimetry at baseline and during the 10th week of intervention. Consumption of fructose, but not glucose, led to significant decreases of net postprandial fat oxidation and significant increases of net postprandial carbohydrate oxidation (P<0.0001 for both). Resting energy expenditure (REE) decreased significantly from baseline values in subjects consuming fructose (P=0.031) but not in those consuming glucose. Increased consumption of fructose for 10 weeks leads to marked changes of postprandial substrate utilization including a significant reduction of net fat oxidation. In addition, we report that REE is reduced compared with baseline values in subjects consuming fructose-sweetened beverages for 10 weeks.

A diet high in protein has been shown to have beneficial effects on weight loss and triglyceride (TG) levels when combined with exercise. Recent research has also shown that a diet high in protein in the absence of exercise promotes more favorable results for individuals above the median TG (mTG) levels (>133 mg/dL). The purpose of Chapter 10, by Oliver and colleagues, was to determine if women with TG above median values experience greater benefits to a diet and circuit resistance-training program.

Black tea is, second only to water, the most consumed beverage globally. Previously, the inhibition of DNA methyltransferase 1 was shown by dietary polyphenols and epi-gallocatechin gallate (EGCG), the main polyphenolic constituent of green tea, and 5-caffeoyl quinic acid, the main phenolic constituent of the green coffee bean. In Chapter 11, Rajavelu and colleagues studied the inhibition of DNA methyltransferase 3a by a series of dietary polyphenols from black tea such as theaflavins and thea-rubigins and chlorogenic acid derivatives from coffee. For theaflavin 3,3 digallate and thearubigins IC50 values in the lower micro molar range were observed, which when compared to pharmacokinetic data available, suggest an effect of physiological relevance. Since Dnnmt3a has been associated with development, cancer and brain function, these data suggest a biochemical mechanism for the beneficial health effect of black tea and coffee and a possible molecular mechanism for the improvement of brain performance and mental health by dietary polyphenols.

Obesity is a common pathology with increasing incidence, and is associated with increased mortality and healthcare costs. Several treatment options for obesity are currently available ranging from behavioral modifications to pharmaceutical agents. Many popular dietary supplements claim to enhance weight loss by acting as metabolic stimulators, however direct tests of their effect on metabolism have not been performed. Chapter 12, by Vaughan and colleagues, identified the effects of popular dietary supplements on metabolic rate and mitochondrial biosynthesis in human skeletal muscle cells. Human rhabdomyosarcoma cells were treated with popular dietary supplements at varied doses for 24 hours. Peroxisome proliferator-activated receptor coactivator 1 alpha (PGC-1α), an important stimulator of mitochondrial biosynthesis, was quantified using quantitative reverse transcriptase polymerase chain reaction (qRT-PCR). Mitochondrial content was measured using flow cytometry confirmed with confocal microscopy. Glycolytic metabolism was quantified by measuring extracellular acidification rate (ECAR) and oxidative metabolism was quantified by measuring oxygen consumption rate (OCR). Total relative metabolism was quantified using WST-1 end point assay. Treatment of human rhabdomyosarcoma cells with dietary supplements OxyElite Pro (OEP) or Cellucore HD (CHD) induced PGC-1α leading to significantly increased mitochondrial content. Glycolytic and oxidative capacities were also significantly increased following treatment with OEP or CHD. This is the first work to identify metabolic adaptations in muscle cells following treatment with popular dietary supplements including enhanced mitochondrial biosynthesis, and glycolytic, oxidative and total metabolism.

PART I

NUTRITIONAL GENOMICS

CHAPTER 1

Bellagio Report on *Healthy Agriculture, Healthy Nutrition, Healthy People*

ARTEMIS P. SIMOPOULOS, PETER G. BOURNE,
AND OLE FAERGEMAN

1.1 INTRODUCTION

The meeting on *Healthy Agriculture, Healthy Nutrition, Healthy People* took place at the Rockefeller Foundation Bellagio Center in Italy, 30 October–1 November 2012. The meeting was sponsored by The Center for Genetics Nutrition and Health, The Rockefeller Foundation, Green Templeton College of the University of Oxford, W.K. Kellogg Foundation, Nutrilite Health Institute, Health Studies Collegium, Hellenic American University, and Hellenic American Union. The focus of the meeting was the Implementation of the Action Plan on Healthy Agriculture, Healthy Nutrition, Healthy People, which had been developed at the meeting on Healthy Agriculture, Healthy Nutrition, Healthy People, held at Ancient Olympia, Greece, 5–8 October 2010 [1,2].

The meeting in Bellagio was science-based but policy-oriented. There were 19 participants from 9 countries, including distinguished physicians, nutritionists, agriculturists, economists, policy experts, lawyers, representatives of industry and representatives from United States Agency for International Development (USAID), Pan American Health Organization (PAHO), and the W.K. Kellogg Foundation. This international constellation of expertise provided a superb opportunity for in depth discussions of the most current scientific evidence on sustainable agriculture and nutrition security for health.

The group's broad, overall concerns were with human health, particularly child health, with societal economics, and with planetary ecosystems. Our lifestyles—including where we live, our activity levels, economic well-being and exposure to stress—all affect human health. We are also embedded in larger systems of agriculture, food cultures and food supply chains that can increase as well as decrease our chances of becoming and remaining healthy. At the same time that some children starve, others (sometimes in the same societies) are prone to obesity and other chronic diseases that stem from poor nutritional content of foods.

While many substances in the diet may affect health, the meeting focused primarily on those elements where the scientific evidence shows the link to be strongest and where the impact on the epidemic of non-communicable diseases (NCDs) worldwide is greatest.

1.2 GOALS

1. To develop strategies that would translate the current state of scientific knowledge on nutrition into specific interventions that will result in people eating healthier diets.
2. Agronomic, nutritional and medical sciences should not be subservient to business interests.

1.3 THE MEETING FOCUSED ON THE FOLLOWING ISSUES

1. Health-oriented agriculture is needed to tailor the food chain to eradicate critical deficiencies and imbalances (e.g., change animal

feeds to balance the omega-6/omega-3 fatty acid ratio, decrease the excessive production of high fructose corn syrup (HFCS) [3,4,5,6].

2. Agronomic, nutritional and medical sciences should be independent of business interests.

3. Need for new forms of agriculture such as agroecology and urban agriculture.

4. Future dietary guidelines to be based on ecological (including climatological) as well as nutritional science.

5. Nutrition research should be the basis of food sciences research and not the reverse as it is now.

1.4 NOVEL ASPECTS OF THE MEETING

Over the past 10 years there have been many reports on diet and chronic diseases, obesity, global health, and non-communicable diseases (NCD's) issued by WHO-FAO, national governments, scientific institutions, medical associations and foundations. However, the novelty of this meeting was the emphasis on:

1. The vital role of political leadership in translating the current well-documented level of scientific knowledge into national and international policies that will change the composition of the food people consume.

2. The importance of specific nutrients such as the balance of omega-6 and omega-3 fatty acids in the diet and the excessive production and consumption of fructose and its detrimental effects on growth and development of children and the development of chronic diseases [7].

3. The importance of developing national food composition tables. To date, only a few countries have these data, and as a result FAO and WHO depend on data per capita consumption of major food groups for policy making [8].

4. The differences and similarities between more affluent countries and developing countries, and what needs to be done going forward that is practical, feasible, and sustainable.

5. The economic and political contexts in which meaningful actions affecting population nutrition must occur.
6. The role of genetics. Gene expression patterns and their frequencies differ geographically between populations and within populations, but the effect of genetic variants on disease is modified by environmental factors including diet. For example, the dietary intake of vegetable oils high in omega-6 fatty acids increases the risk for cardiovascular disease as a function of genetic variation in European populations and perhaps even more so in populations of African ancestry with genetic variants affecting rates of metabolism of omega-6 fatty acids due to their higher frequency [9,10]. Gene/diet interactions should be considered in all studies relating diet to health and diseases such as diabetes, obesity, cardiovascular disease and African sleeping sickness. Recently the NIH and Wellcome Trust have joined forces to fund large-scale population studies by African researchers on African populations.

1.5 MEETING CONTENT

One of the important issues included in all the presentations, given over three days, was the obstacles governments face in implementing policies that would lead to optimal scientifically based diets for their populations. The sources of opposition to those policies were also discussed. Participants were asked to consider the complexities for government adoption of policy, including regional considerations, leadership issues (academic institutions, medical school education, industry, including agribusiness); management; economic issues (which was discussed in detail by Dr. Ole Faergeman of Denmark); and nutritional determinants of health. Specific attention was also given to commercial considerations with attention to developing policies without unnecessary negative impact on the food industry. However, it was agreed government must not be influenced by industry to pursue policies contrary to the health and nutrition needs of its people.

1.6 LOCAL INITIATIVES TO ENLIGHTEN INDUSTRY

Two countries have initiated nutritional programs with demonstrable benefit on the nutrition and health of their peoples. In both instances a pre-condition for success was a high level of political commitment to assuring a strong food supply with optimal nutritional content. Dr. Dan L. Waitzberg (Brazil) gave a presentation of how Brazilian government policy, under presidential direction, has resulted in the "right to nutrition and food" for all of its citizens, and how this new policy has had an impact on the health of the Brazilian people [11]. Similarly, Dr. Kraisid Tontisirin (Thailand) provided an exciting presentation on how the nutrition, agriculture and health departments of the Thai government have worked together to develop nutritional policies based on consideration of all three disciplines [12].

1.7 ROLE OF SPECIFIC FOOD GROUPS

One of the more hotly discussed topics was on the role of the nutritional content of foods. This included a discussion of the role of healthy and unhealthy fats, with attention to the relative content of omega-3 and omega-6 fatty acids in food by Dr. Artemis P. Simopoulos (USA) [13,14,15,16,17]. In addition, Dr. Richard J. Johnson (USA) reviewed the evidence that the worldwide increase in added sugars containing fructose may underlie the epidemics of obesity and diabetes [7,18,19,20].

1.8 CONCLUSIONS

1.8.1 GENERAL CONCLUSIONS

1. Good health requires food of good quality. Access to optimal nutrition and health are fundamental human rights. They apply to us all, rich and poor, young and old.
2. Malnutrition remains common. One in seven humans is malnourished because of poverty. The poor live in poor countries, but they

also live in rich countries with major inequalities in wealth. The poor have little choice about what is available for them to eat. In contrast, the affluent suffer from overnutrition having a wide selection of foods with both poor and good nutritional content, but inadequate knowledge or government guidance to avoid a diet that impacts adversely on their health.

3. Malnutrition is a societal issue, and it is a gigantic one. The last 30 years have seen a dramatic growth of economic and geopolitical power of emerging markets—Brazil, China, India, Indonesia, Mexico, Russia, South Africa—and the magnitude of the nutrition and health issues facing those countries will soon exceed that in wealthy countries. The issues include simultaneously both the continuously growing chronic non-communicable diseases of the affluent, and the infectious diseases of the poor. The dietary choices made by the affluent in these countries will have a steadily increasing negative impact on the health of their populations. There is an opportunity to prevent this.

4. Malnutrition is also a function of what food we choose to produce, how we produce it, and whether and how we make it available to us all. Farmers, industrial agriculture and food processing and distribution profoundly affect ecosystems and climate, moreover, and they are major actors in our economic and financial systems. These players also directly determine the quality of the dietary options available.

5. Good nutrition and malnutrition are understood by scientists, and they should have a key role in the adoption of good nutrition, ecology and agronomy policies by governments.

6. Governments are influenced to varying degrees by corporate interests. The task of government leaders is to work out policies for food and nutrition with appropriate respect to culture and agricultural tradition as well as the food industry.

7. Adding to the difficulties of formulating policy, the ivory tower of unadulterated university research no longer exists. The food industry including agriculture also performs research, it contributes to and influences research performed in universities, and it understandably exploits the results of research in choosing what

to produce and bring to market. All of these various complexities continue to affect if not determine debates about nutrition and human health.

8. Chronic non-communicable diseases such as atherosclerosis, type 2 diabetes, obesity, respiratory diseases, and certain cancers are common in rich countries and on the increase in countries on the way to affluence. All of these conditions are more or less determined by what we eat, and the debates about what to eat to avoid disease are almost countless. Some of them are nevertheless scientifically well informed by studies at many different levels of nutrition understanding.

9. Well performed epidemiological studies have documented effects of micronutrients on health. Vitamin D is an example. Still other studies have been performed on the level of the three major sources of food energy: carbohydrates, fats and proteins. The following discussion is an attempt to exemplify aspects of this particular branch of nutrition research that are relevant for policy.

10. Advancing science has provided convincing evidence that providing food solely based on calorie content is not enough to provide good health and nutrition. Rather, the choice of carbohydrates, fats, and proteins affect risk of disease. Even obesity is not a simple function of caloric intake. Increasing evidence, for example, suggests that large amounts of a sugar such as fructose in processed foods and beverages may increase the risk for developing diabetes and liver disease. The adverse effects of excessive sugar consumption had been known for more than fifty years but we have failed to intervene appropriately.

1.8.2 SPECIFIC CONCLUSIONS

1.8.2.1 FRUCTOSE FROM ADDED SUGARS

1. Fructose is a monosaccharide found in honey, ripened fruits and vegetables. Table sugar is sucrose, a disaccharide composed of fructose chemically coupled to glucose, another monosaccharide.

Sources of fructose are sugar cane, sugar beets and corn. It is an effective and low-cost sweetener, and it is therefore extensively used in food and beverages (high-fructose corn syrup, HFCS). It does have a dependence producing effect making it hard for people to reduce or eliminate it from their diet [21,22,23].

2. Intake of sugar and sweeteners containing fructose has increased markedly in many countries throughout the world. The U.S. National Health and Nutrition Examination Survey (NHANES), for example, reported that about 15% of Americans consume greater than 25% of their energy from added sugars. Annual intake of added sugars in the United States is approximately 35 kg/capita or about one sixth of food energy.

3. There is increasing evidence from experimental and clinical studies that intake of added sugars not only increases the well-known risk of caries, but also risk of cardiovascular disease, non-alcoholic fatty liver disease, obesity, diabetes, and possibly even cancer. While some authorities, primarily those funded by the food industry, have argued that the high amounts of added sugars in food and beverages may contribute to health risks solely as a consequence of their caloric content, there is also mounting evidence that fructose may have a specific ability to cause fatty liver (which can progress to cirrhosis of the liver), high triglycerides in blood (which can contribute to cardiovascular disease), insulin resistance (leading to type 2 diabetes) and increased appetite (which obviously can lead to obesity) [5,6]. Obesity itself promotes cardiovascular disease, type 2 diabetes and certain cancers, moreover. Immoderate intake of added sugars, fructose in particular, may therefore increase health risks with important public health implications.

1.8.2.2 FATTY ACIDS

1. Studies performed since the middle of the 20th century indicated that saturated fat increased, and polyunsaturated fat lowered, the risk of disease, cardiovascular disease in particular. That understanding encouraged farmers and the food industry to increase

production of vegetable oils rich in polyunsaturated fats from soy, sunflower and, particularly in the United States, corn (maize).

2. Fat in food is mainly fatty acids chemically coupled to glycerol. Fatty acids can be saturated with hydrogen. If not, they are more or less unsaturated. The polyunsaturated fatty acids contribute importantly to average diets, but the balance of two kinds of polyunsaturated fatty acids in modern diets is quite different from that in diets during human evolution [3,16]. Whereas the latter contained about one omega-3 fatty acid for every four omega-6 fatty acids, modern diets can contain as much as fifty to a hundred times more omega-6 than omega-3 polyunsaturated fatty acids. The evidence that this imbalance contributes to disease is now convincing, and governments should formulate policies for agriculture and food to affect costs and availability of various fatty acids to the general public so that the ratio of omega-6 to omega-3 fatty acids can once again approach that to which we are genetically adapted, i.e. four to one [4,24]. High omega-6/omega-3 ratios typify Western diets and, increasingly, diets throughout the world, and they are associated with increased risk for cardiovascular disease, obesity, type 2 diabetes and cancer of the breast and prostate, particularly in individuals who are genetically predisposed. Of concern, animal experiments indicate that low intakes of docosahexanoic acid, an omega-3 fatty acid, in combination with a high intake of fructose, leads to metabolic syndrome in the brain [25].

1.8.2.3 ALL CALORIES ARE NOT THE SAME

1. We use the apparent self-contradiction, "a calorie is not a calorie", to emphasize that different nutrients with the same amount of food energy (calories) can differ in their effects on body weight. Fructose, for example, increases appetite more effectively than glucose [6,26]. One calorie of fructose is therefore more obesogenic than one calorie of glucose. Similarly, omega-6 fatty acids may be more obesogenic than omega-3 fatty acids. Weight loss regimens must therefore take nutritional as well as overall caloric concerns into account [27,28].

2. The metabolic effects of whole food calories also differ from those of processed and restructured foods [29].

1.8.2.4 NUTRITION IS PART OF A LARGER PICTURE

1. We acknowledge the monetary importance of agriculture and food production, but we also acknowledge the importance of agriculture for societal fabric and the impact of agriculture on the ecosystems on which we depend. World-wide increased agricultural production is ascribed to the further industrialization of agriculture since the mid 20th century ("Green Revolution"), but industrial agriculture is also an important reason that mankind has now passed several planetary boundaries for sustainability.

2. They include disruption of the nitrogen cycle, loss of biodiversity, and global warming. Demand for chemical fertilizers is also rapidly depleting known deposits of phosphorus, and profligate use of phosphorus, nitrogen and pesticides is an important cause of destruction of ecosystems including those in soil. Others compromising soil health are erosion by wind and water, compaction by heavy machinery, and pollution by effluents from intensive production of livestock.

3. The allocation of farm land to raising biofuels and feed for animals rather than food for humans increases demand for and transnational purchases of farm land in poor countries by rich countries. Such allocation also increases the price of food. Food prices have fluctuated, moreover, because of the speculation in agricultural commodities made possible by deregulation of financial markets. Most of these complexities affect nutrition detrimentally, and they all make life more difficult and precarious for the poor.

1.8.2.5 THE BRAZILIAN MODEL

1. Brazil is a good example of how presidential leadership can mobilize all aspects of government, national agriculture and public

health to achieve better health through dramatically improved nutrition. President Luiz Ignacio Lula da Silva publicly announced the very high priority he attached to ending hunger and reducing poverty in the country. This set in motion changes throughout the society that enhanced the availability and nutritional quality of food. The government provided leadership in supporting local food production. Legislation required that 30% of meals served at schools must come from local markets, thus supporting local farmers and providing fresh and nutritious foods consistent with the culture of the various local communities.

2. Brazil has sought original ways to eliminate hunger and poverty, obliging the state to implement public policies that guarantee fundamental human rights to minimum income, food, health, education and work.

3. Some of the key lessons learned include: (i) the importance of participatory pacts related to concepts and principles; (ii) the appropriateness of the choice of a systemic and intersectoral approach; (iii) the relevant role of civil society ensured through formal spaces of social dialogue (CONSEAs); (iv) the importance of the state in the protection of human rights above market interests; (v) the necessary practice of intersectoral coordination in the design and management of public policies on food and nutrition security; (vi) the strategic role of women in the struggle to guarantee food sovereignty as well as the conservation and sustainable management of natural resources; and (vii) the respect for and guarantee of ethno-development principles in the design and implementation of public policies for indigenous peoples, blacks, traditional peoples and communities.

4. The continuity of the main public policies that have contributed to this progress and the convergence of political and social forces are indispensable conditions to overcoming the challenges that still hinder the elimination of all forms of social inequality and violation of rights.

5. Brazil provides a model that other countries can emulate.

1.8.2.6 THE THAI MODEL

1. Thailand, an emerging economy with a distinctive heritage of a unique cultural cuisine, is today one of the leaders in the progressive management of food production, marketing, nutrition and human health. Under the nation's *Strategic Framework for Food Management*, nutrition policy is formulated in a way that goes well beyond the office of any one department or ministry of government. As in Brazil Thailand has shown high level government commitment to food and nutrition policy. Ministers and secretaries of health, food, agriculture, urban and rural planning, commerce, foreign aid, and finance all have responsibilities and rights to guide policy agendas with implications for food. Food policy has therefore been elevated to the highest levels, and the head of government serves as chairperson of the strategic planning group. Perhaps even more progressive is the assumption that the highest levels of government are there to support self-directed community-based leaders in fulfilling locally defined objectives for food production, health promotion and environmental stewardship.

1.9 RECOMMENDATIONS

1. Heads of state and government must elevate, as a matter of urgency, nutrition as a national priority (e.g., Brazil and Thailand).
2. Good nutrition is a human right, but it is impossible to achieve for whole populations without good policies for food, health, nutrition, agriculture, ecology, economy and commerce. It is therefore the responsibility of heads of state and government to provide the leadership that will lead to an "all society" approach for good nutrition.
3. Advance public understanding of the following key aspects of nutrition:

    • With the increasing decline in infectious diseases most experts believe that poor nutrition is now the single most important obstacle to better health worldwide.

- Under-nutrition and malnutrition primarily affect the developing world where people with no choice have inadequate intake of calories and micronutrients. It differs from the problem in industrialized nations where many people knowingly and unknowingly choose a diet with a composition that leads to serious chronic disease and premature death.
- Emerging market countries such as Brazil, China, India and Russia simultaneously face the nutritional problems of both developed and developing countries.
- Sugar consumption, especially in the form of high energy fructose in soft drinks, poses a major and insidious health threat, especially to children. The health threat is comparable to that from cigarette smoking.

   - Most diets, although with regional differences, are deficient in omega-3 fatty acids and too high in omega-6 fatty acids.
   - Access to a healthy diet should be considered a human right.

4. Place the lead responsibility for nutrition in ministries of health rather than agriculture so that the health requirements drive agricultural priorities not vice versa. Nutritional security should be given the same priority as food security.

5. The American Heart Association warnings on the "overconsumption" of added sugar should be strongly promoted (no more than 6 teaspoons for an adult woman and 9 teaspoons for adult men daily) [30]. As an example, this would limit the average woman to one 8-oz sugar-sweetened beverage per day or its equivalent. Health warnings on all sugar-sweetened beverages should be considered.

6. A concerted effort is needed to decrease the ratio of omega-6 to omega-3 fatty acids in the diet. Education and if necessary government intervention should be used to get populations to switch from oils high in omega-6 such as corn, safflower, and sunflower oils, to those high in omega-3 such as rapeseed, flax seed and oils high in monounsaturated fatty acids such as olive oil, hazelnut oil in combination with rapeseed oil. Increased fish consumption should be stressed. Scientists should collaborate with the fishing industry to achieve this end. A ratio of 4:1 of omega-6 to omega-3 in the diet should be the goal.

7. Governments through their agricultural policies, taxation, subsidies, pricing and controls at the point of distribution should support the availability of foods rich in healthful components. They should also strongly consider penalizing those who put on the market products that are harmful to health. In so doing governments should place a higher emphasis on the health of the population over market interests. They should also foster and support cultivation at the local level including urban agriculture. The production of vegetables and fruits high in anti-oxidants should be stressed.

8. In view of the limited knowledge most physicians and other health providers have concerning nutrition, a major initiative should be launched to incorporate nutrition into curricula stressing its crucial role in the epidemic of non-communicable diseases. A similar initiative should be launched with those already practicing.

9. Food consumption patterns vary around the globe as a result of food availability, cultural determinants, and economic circumstances. A series of Research Centers on Genetics, Nutrition and Fitness for Health should be established in different regions, along with educational components for professionals and the public. They would collect and analyze food consumption data focusing particularly on the chemical content of the food consumed in their regions.

10. The Center for Genetics Nutrition and Health representing the Bellagio group will work to implement the conclusions reached at the meeting of 29 October–2 November 2012. This will include:

- Distributing copies of the Bellagio Report to a wide diversity of academic and non-academic outlets.
- Consulting and collaborating with other groups concerned with malnutrition and under-nutrition, food security, access to a nutritionally adequate diet for the economically deprived, and particularly the health and nutrition of children.
- Working with leaders of national governments, both executive and legislative, and international organizations such as FAO, WHO and its regional organizations, especially PAHO, the World Bank and other agencies of the UN to achieve the incorporation of the recommendations above into their policies and programs.
- Establishing a coalition with the WHO Commissions on Non-Communicable Disease and the Social Determinants of Health to insure that the essentiality of nutrition for normal growth and development

and in combating chronic non-communicable diseases is always considered in their deliberations.

REFERENCES

11. Simopoulos, A.P. Healthy Agriculture, Healthy Nutrition, Healthy People; World Review of Nutrition and Dietetics Volume 102; Karger: Basel, Switzerland, 2011.
12. Simopoulos, A.P.; Faergeman, O.; Bourne, P.G. Action Plan for a Healthy Agriculture, Healthy Nutrition, Healthy People. J. Nutrigenet. Nutrigenomics 2011, 4, 65–82, doi:10.1159/000328438.
13. Simopoulos, A.P. The importance of the omega-6/omega-3 Fatty Acid ratio in cardiovascular disease and other chronic diseases. Exp. Biol. Med. (Maywood) 2008, 233, 674–688, doi:10.3181/0711-MR-311.
14. Simopoulos, A.P. Omega-6/Omega-3 Essential Fatty Acids: Biological Effects. In Omega-3 Fatty Acids, the Brain and Retina; World Review of Nutrition and Dietetics Volume 99; Simopoulos, A.P., Bazan, N.G., Eds.; Karger: Basel, Switzerland, 2009; pp. 1–16.
15. Stanhope, K.L.; Schwarz, J.M.; Keim, N.L.; Griffen, S.C.; Bremer, A.A.; Graham, J.L.; Hatcher, B.; Cox, C.L.; Dyachenko, A.; Zhang, W.; et al. Consuming fructose-sweetened, not glucose-sweetened, beverages increases visceral adiposity and lipids and decreases insulin sensitivity in overweight/obese humans. J. Clin. Invest. 2009, 119, 1322–1334, doi:10.1172/JCI37385.
16. Stanhope, K.L.; Bremer, A.A.; Medici, V.; Nakajima, K.; Ito, Y.; Nakano, T.; Chen, G.; Fong, T.H.; Lee, V.; Menorca, R.I.; Keim, N.L.; Havel, P.J. Consumption of fructose and high fructose corn syrup increase postprandial triglycerides, LDL-cholesterol, and apolipoprotein-B in young men and women. J. Clin. Endocrinol. Metab. 2011, 96, E1596–E1605, doi:10.1210/jc.2011-1251.
17. Te Morenga, L.; Mallard, S.; Mann, J. Dietary sugars and body weight: Systematic review and meta-analyses of randomised controlled trials and cohort studies. BMJ 2012, 346, e7492.
18. Simopoulos, A.P.; Butrum, R.R. International Food Data Bases and Information Exchange: Concepts, Principles and Designs; World Review of Nutrition and Dietetics Volume 68; Karger: Basel, Switzerland, 1992.
19. Sergeant, S.; Hugenschmidt, C.E.; Rudock, M.E.; Ziegler, J.T.; Ivester, P.; Ainsworth, H.C.; Vaidya, D.; Case, L.D.; Langefeld, C.D.; Freedman, B.I.; Bowden, D.W.; Mathias, R.A.; Chilton, F.H. Differences in arachidonic acid levels and fatty acid desaturase (FADS) gene variants in African Americans and European Americans with diabetes or the metabolic syndrome. Br. J. Nutr. 2012, 107, 547–555.
20. Mathias, R.A.; Sergeant, S.; Ruczinski, I.; Torgerson, D.G.; Hugenschmidt, C.E.; Kubala, M.; Vaidya, D.; Suktitipat, B.; Ziegler, J.T.; Ivester, P.; et al. The impact of FADS genetic variants on ω6 polyunsaturated fatty acid metabolism in African Americans. BMC Genet. 2011, 12, 50.

21. Leao, M.; Maluf, R.S. Effective Public Policies and Active Citizenship: Brazil's Experience of Building a Food and Nutrition Security System; Ação Brasileira pela Nutrição e Direitos Humanos (ABRANDH): Brasília, Brazil, 2012; p. 73.
22. The Thailand Food Committee. Strategic Framework for Food Management in Thailand. Available online: http://www.tnfc.in.th (accessed on 12 August 2012).
23. Daak, A.A.; Ghebremeskel, K.; Hassan, Z.; Attallah, B.; Azan, H.H.; Elbashir, M.I.; Crawford, M. Effect of omega-3 (n-3) fatty acid supplementation in patients with sickle cell anemia: Randomized, double-blind, placebo-controlled trial. Am. J. Clin. Nutr. 2013, 97, 37–44, doi:10.3945/ajcn.112.036319.
24. Skilton, M.R.; Mikkilä, V.; Würtz, P.; Ala-Korpela, M.; Sim, K.A.; Soininen, P.; Kangas, A.J.; Viikari, J.S.; Juonala, M.; Laitinen, T.; Lehtimäki, T.; Taittonen, L.; Kähönen, M.; Celermajer, D.S.; Raitakari, O.T. Fetal growth, omega-3 (n-3) fatty acids, and progression of subclinical atherosclerosis: preventing fetal origins of disease? The Cardiovascular Risk in Young Finns Study. Am. J. Clin Nutr. 2013, 97, 58–65, doi:10.3945/ajcn.112.044198.
25. Li, J.; Xun, P.; Zamora, D.; Sood, A.; Liu, K.; Daviglus, M.; Iribarren, C.; Jacobs, D., Jr.; Shikany, J.M.; He, K. Intakes of long-chain omega-3 (n-3) PUFAs and fish in relation to incidence of asthma among American young adults: the CARDIA study. Am. J. Clin. Nutr. 2013, 97, 173–178, doi:10.3945/ajcn.112.041145.
26. Gibson, R.A.; Neumann, M.A.; Lien, E.L.; Boyd, K.A.; Tu, W.C. Docosahexaenoic acid synthesis from alpha-linolenic acid is inhibited by diets high in polyunsaturated fatty acids. Prostaglandins Leukot. Essent. Fatty Acids. 2013, 88, 139–146, doi:10.1016/j.plefa.2012.04.003.
27. Ramsden, C.E.; Hibbeln, J.R.; Majchrzak, S.F.; Davis, J.M. n-6 fatty acid-specific and mixed polyunsaturate dietary interventions have different effects on CHD risk: a meta-analysis of randomised controlled trials. Br. J. Nutr. 2010, 104, 1586–1600, doi:10.1017/S0007114510004010.
28. Johnson, R.J.; Segal, M.S.; Sautin, Y.; Nakagawa, T.; Feig, D.I.; Kang, D.H.; Gersch, M.S.; Benner, S.; Sánchez-Lozada, L.G. Potential role of sugar (fructose) in the epidemic of hypertension, obesity and the metabolic syndrome, diabetes, kidney disease, and cardiovascular disease. Am. J. Clin. Nutr. 2007, 86, 899–906.
29. Page, K.A.; Chan, O.; Arora, J.; Belfort-Deaguiar, R.; Dzuira, J.; Roehmholdt, B.; Cline, G.W.; Naik, S.; Sinha, R.; Constable, R.T.; Sherwin, R.S. Effects of fructose vs. glucose on regional cerebral blood flow in brain regions involved with appetite and reward pathways. JAMA 2013, 309, 63–70.
30. Purnell, J.Q.; Fair, D.A. Fructose ingestion and cerebral, metabolic, and satiety responses. JAMA 2013, 309, 85–86, doi:10.1001/jama.2012.190505.
31. Handbook of Food and Addiction; Brownell, K., Gold, M., Eds.; Oxford University Press: New York, NY, USA, 2012.
32. Green, S.M.; Blundell, J.E. Effect of fat- and sucrose-containing foods on the size of eating episodes and energy intake in lean dietary restrained and unrestrained females: potential for causing overconsumption. Eur. J. Clin. Nutr. 1996, 50, 625–635.
33. Green, S.M.; Burley, V.J.; Blundell, J.E. Effect of fat- and sucrose-containing foods on the size of eating episodes and energy intake in lean males: potential for causing overconsumption. Eur. J. Clin. Nutr. 1994, 48, 547–555.

34. Simopoulos, A.P.; Cleland, L.G. Omega-6/Omega-3 Essential Fatty Acid Ratio: The Scientific Evidence; World Review of Nutrition and Dietetics Volume 92; Karger: Basel, Switzerland, 2003.

35. Agrawal, R.; Gomez-Pinilla, F. "Metabolic syndrome" in the brain: deficiency in omega-3 fatty acid exacerbates dysfunctions in insulin receptor signalling and cognition. J. Physiol. 2012, 590, 2485–2499, doi:10.1113/jphysiol.2012.230078.

36. Ishimoto, T.; Lanaspa, M.A.; Le, M.T.; Garcia, G.E.; Diggle, C.P.; Maclean, P.S.; Jackman, M.R.; Asipu, A.; Roncal-Jimenez, C.A.; Kosugi, T.; et al. Opposing effects of fructokinase C and A isoforms on fructose-induced metabolic syndrome in mice. Proc. Natl. Acad. Sci. USA 2012, 109, 4320–4325.

37. Moon, R.J.; Harvey, N.C.; Robinson, S.M.; Ntani, G.; Davies, J.H.; Inskip, H.M.; Godfrey, K.M.; Dennison, E.M.; Calder, P.C.; Cooper, C. The SWS Study Group. Maternal Plasma Polyunsaturated Fatty Acid Status in Late Pregnancy Is Associated with Offspring Body Composition in Childhood. J. Clin. Endocrinol. Metab. 2013, 98, 299–307, doi:10.1210/jc.2012-2482.

38. Massiera, F.; Saint-Marc, P.; Seydoux, J.; Murata, T.; Kobayashi, T.; Narumiya, S.; Guesnet, P.; Amri, E.Z.; Negrel, R.; Ailhaud, G. Arachidonic acid and prostacyclin signaling promote adipose tissue development: a human health concern? J. Lipid Res. 2003, 44, 271–279, doi:10.1194/jlr.M200346-JLR200.

39. Ludwig, D.S. The glycemic index: physiological mechanisms relating to obesity, diabetes, and cardiovascular disease. JAMA 2002, 287, 2414–2423, doi:10.1001/jama.287.18.2414.

40. Johnson, R.K.; Appel, L.J.; Brands, M.; Howard, B.V.; Lefevre, M.; Lustig, R.H.; Sacks, F.; Steffen, L.M.; Wylie-Rosett, J. American Heart Association Nutrition Committee of the Council on Nutrition, Physical Activity, and Metabolism and the Council on Epidemiology and Prevention. Dietary sugars intake and cardiovascular health: a scientific statement from the American Heart Association. Circulation 2009, 120, 1011–1020.

CHAPTER 2

Diet, Genetics, and Disease: A Focus on the Middle East and North Africa Region

AKL C. FAHED, ABDUL-KARIM M. EL-HAGE-SLEIMAN, THERESA I. FARHAT, AND GEORGES M. NEMER

2.1 INTRODUCTION

Over the past few decades, the MENA has been witnessing significant changes in food habits paralleled by an important preponderance of metabolite-related diseases. In a region whose traditional diet is known to be healthy due to high vegetable proteins, fibers, minerals, and vitamins with low content of unfavorable food products, the "industrialization/westernization of the diet" is a well-studied and documented phenomenon [1–3]. The MENA has been losing its traditional diet which was distinguished by its diversity and richness in raw foods, proteins, and multivitamins, in the favor of a more industrial diet which consists of increased preprocessed foods, sugars, fats, alcohol, animal products, saturated- and trans-fatty acids, and relatively less vitamins and minerals with decreased consumption

Diet, Genetics, and Disease: A Focus on the Middle East and North Africa Region. © Fahed AC, El-Hage-Sleiman A-KM, Farhat TI, and Nemer GM. Journal of Nutrition and Metabolism *2012 (2012).* http://dx.doi.org/10.1155/2012/109037. *Licensed under Creative Commons Attribution 3.0 Unported License, http://creativecommons.org/licenses/by/3.0/.*

of milk, fruits, and vegetables [4]. A big part of this change is attributed to the lifestyle changes and globalization with the invasion of western fast food to the MENA countries. Dietary choices, minimum physical activity, religious habits, consumer ignorance, high population growth rates, economic factors, and lack of both protection laws and food fortification programs are other critical factors that influence the nutritional status in the region [5]. These changes in dietary and lifestyle patterns contribute to an increase in the rates of micronutrients deficiencies, diet-related chronic diseases, and obesity in all groups of the population in the region [5]. Due to this grave impact on chronic diseases, diet became a target of public health initiatives that aim at restoring the traditional diet of MENA countries in order to improve health conditions in their populations [6–8]. The epidemiology of diet-related diseases in the MENA region and background information on the diet-genetics-disease interaction, followed by nutrigenomic examples on diet-related diseases in the MENA region, will be discussed for the first time in this paper.

1.2 EPIDEMIOLOGY OF DIET-RELATED DISEASES IN THE MENA

We review the numbers and trends for selected chronic metabolic diseases and micronutrient deficiencies in the different countries of the MENA where data are available (Figure 1). Given the lack of nationwide data, prevalence rates are commonly reported as estimations [9, 10]. Since 1982, the need for "direct evidence of a secular [increase]" in diseases has been established, prospecting an association with "acculturation" of traditional or rural populations [11]. Still, the numbers show alarming trends for cardiovascular diseases and metabolic disorders, namely, insulin resistance, adiposity, dyslipidemias, and atherosclerosis. Likewise, micronutrient deficiencies (MNDs) have been heavily studied recently due to their crucial contribution to the global burden of many chronic diseases.

The region is witnessing an "explosion" of Type 2 Diabetes Mellitus (T2DM) according to the International Diabetes Foundation (IDF) (Figure 2(a)) with five countries of the MENA ranking among the top ten diabetic worldwide [15]. Similarly, the IDF reports an increase in the incidence of Type 1 DM in children of the MENA over the past decade (Figure 2(b)).

The World Health Statistics of the WHO show a sharp rise in the prevalence of obesity (Figure 3), which is the most powerful and easily reported risk marker and hence the sole epidemiological WHO indicator for cardiovascular and metabolic comorbidities. In one epidemiological study, cardiovascular mortality in the MENA has been estimated to triple from 1990 to 2020 (Figure 4). Table 1 summarizes rates of cardiovascular diseases and two major comorbid risk factors, hypertension and the metabolic syndrome, with wide variations across countries of the MENA.

TABLE 1: Rates of cardiovascular disease, hypertension, and the metabolic syndrome in MENA countries from different studies.

Coronary artery disease (CAD)			
Iran	Age-adjusted prevalence (%)	12.7	Nabipour et al. [73]
Jordan	Prevalence (%)	5.9	Nsour et al. [74]
Saudi Arabia,			
(rural)	Prevalence (%)	4.0	Al-Nozha et al. [75]
(urban)	Prevalence (%)	6.2	
(overall)	Prevalence (%)	5.5	
Tunisia	Prevalence (%), [men]	12.5	Ben Romdhane et al. [76]
	Prevalence (%), [women]	20.6	
Cerebrovascular Accidents			
Bahrain	Age-adjusted incidence (per 100,000)	96.2	Al-Jishi and Mohan [77]
Iran	Age-adjusted incidence (per 100,000)	61.5	Ahangar et al. [78]
Kuwait	Age-adjusted incidence (per 100,000)	92.2	Abdul-Ghaffar et al. [79]
Libya	Age-adjusted incidence (per 100,000)	114.2	Radhakrishnan et al. [80]
Palestine	Age-adjusted incidence (per 100,000)	62.7	Sweileh et al. [81]
Qatar	Age-adjusted incidence (per 100,000)	123.7	Hamad et al. [82]
Saudi Arabia	Age-adjusted incidence (per 100,000)	38.5	Al-Rajeh et al. [83]
Hypertension (HTN)			
Algeria	Prevalence (%), [Age > 25]	36.2	Yahia-Berrouiguet et al. [84]
Bahrain	Prevalence (%), [Age > 20]	42.1	Al-Zurba [85]
Egypt	Age-adjusted prevalence (%)	27.4	Ibrahim et al. [86]
	Prevalence (%), [Age > 25]	26.3	Galal [87]
Iran	Prevalence (%), [Age > 19]	25.6	Sarraf-Zadegan et al. [88]
	Prevalence (%), [Age: 30–55]	23.0	Haghdoost et al. [89]
	Prevalence (%), [Age > 55]	49.5	

TABLE 1: *Cont.*

Iraq	Prevalence (%), [Age > 20]	19.3	WHO: STEPwise, [90]
Jordan	Prevalence (%), [Age > 18]	30.2	Zindah et al. [91]
Lebanon	Prevalence (%), [Age: 18–65]	31.2	Sibai et al. [92]
Morocco	Prevalence (%), [Age > 20]	33.6	Tazi et al. [93]
Oman	Prevalence (%), [Age > 20]	21.5	Hasab et al. [94]
Palestine (WB),			
(rural)	Prevalence (%), [Age: 30–65]	25.4	Abdul-Rahim et al. [95]
(urban)	Prevalence (%), [Age: 30–65]	21.5	
Qatar	Prevalence (%), [Age: 25–65]	32.1	Bener et al. [96]
Saudi Arabia	Prevalence (%), [Age: 30–70]	26.1	Al-Nozha et al. [97]
Sudan	Prevalence (%), [Age: 25–64]	23.6	WHO: STEPwise, [90]
Syria	Prevalence (%), [Age: 18–65]	40.6	Maziak et al. [98]
Turkey	Age-adjusted prevalence (%)	25.7	Sonmez et al. [99]
UAE	Prevalence (%), [Age > 20]	20.8	Baynouna et al. [100]
Yemen	Prevalence (%), [Age > 35]	26.0	Gunaid and Assabri [101]
Middle East	Prevalence (%), (overall)	21.7	Motlagh et al. [102]
Metabolic Syndrome			
Algeria	Prevalence (%), [Age > 20]	17.4	Mehio Sibai et al. [103]
Iran	Prevalence (%), [Age > 19]	23.3	Mehio Sibai et al. [103]
Jordan	Prevalence (%), [Age > 18]	36.3	Khader et al. [104]
Kuwait	Prevalence (%), [Age > 20]	24.8	Al Rashdan and Al Nesef [105]
Lebanon	Prevalence (%), [Age: 18–65]	25.4	Mehio Sibai et al. [103]
Morocco, (rural)	Prevalence (%), [women]	16.3	Rguibi and Belahsen [106]
Oman,			
(overall)	Prevalence (%), [Age > 20]	21.0	Al-Lawati et al. [107]
(Nizwa)	Age adjusted prevalence (%)	8.0	Al-Lawati et al. [107]
Palestine (WB)	Prevalence (%), [Age: 30–65]	17.0	Abdul-Rahim et al. [95]
Qatar	Prevalence (%), [Age > 20]	27.7	Musallam et al. [108]
	Age adjusted prevalence (%)	26.5	Bener et al. [109]
Saudi Arabia	Age adjusted prevalence (%)	39.3	Al-Nozha et al. [110]
Tunisia	Prevalence (%), [Age > 20]	16.3	Bouguerra et al. [111]
UAE	Prevalence (%)	39.6	Malik and Razig [112]

Nonadjusted rates from different studies are not valid for comparison but displayed to present the burden of the morbidities. HTN is defined as BP > 140/90 or use of antihypertensive medications. Metabolic Syndrome definition is based on Adult Treatment Panel III, except for Palestine and Tunisia where, respectively, WHO criteria and hypercholesterolemia (Total Cholesterol ≥5.2 mmol/l) instead of low HDL cholesterol were used. UAE: United Arab Emirates; WB: West Bank [103, 113–116].

FIGURE 1: Map of the Middle East and North Africa (MENA) region. The MENA region includes countries such as Algeria, Armenia, and Turkey, that are not members of the WHO Eastern Mediterranean Region (EMR) that is referred to in the literature.

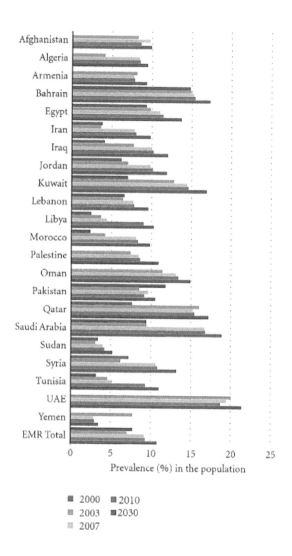

FIGURE 2: Increasing prevalence of diabetes mellitus in the MENA. (a) General increase in reported prevalence (%) of Type 2 Diabetes Mellitus in the MENA between the years 2000 and 2010. Numbers are reported as approximated by the International Diabetes Federation [12–15]. The expected 2-fold increase for the year 2030 is approximated based on demographic parameters, without accounting for changes in age strata or other risk factors [16]. (b) Overall growth in annual incidence (per 100,000) of Type 1 diabetes mellitus in children younger than 14 years old in the MENA. Numbers are estimations by the International Diabetes Federation based on various years between 1986 and 2000 [12–15].

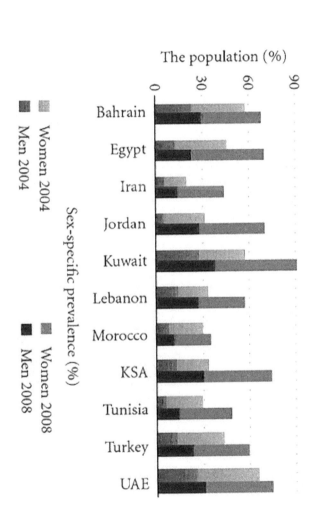

FIGURE 3: Increasing prevalence of obesity in the MENA. Prevalence (%) of obesity increased in both men and women in countries of the MENA between 2004 and 2008. Numbers are WHO estimates in World Health Statistics of 2005 and 2011. Totals of men and women are integrated for purposes of comparative illustration and do not represent adjusted arithmetic total prevalences. Obesity was defined as body mass index (BMI) ≥30 Kg/m². Obesity data about Jordanian men in 2004 are not available, but prevalence was estimated to be less than that of women [17, 18]. KSA: Kingdom of Saudi Arabia; UAE: United Arab Emirates.

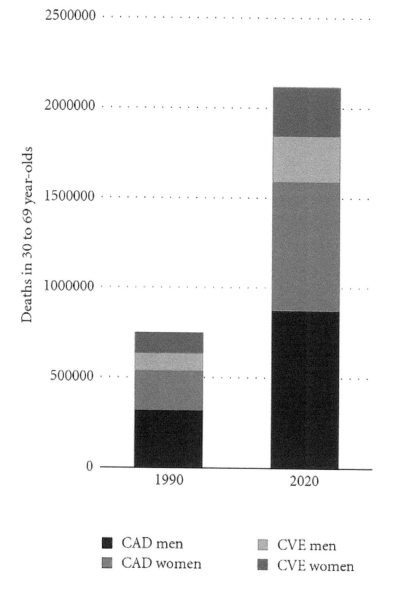

FIGURE 4: Expected overall increase in mortality due to cardiovascular diseases in the MENA [19]. CAD: Coronary Artery Disease; CVE: Cerebrovascular Event.

TABLE 2: Rates of Vitamin D deficiency and iron deficiency in MENA countries from different studies.

Vitamin D deficiency (VDD)			
Iran	Prevalence (%), [girls], [adolescent]	Up to 70	Moussavi et al. [143]
Jordan	Prevalence (%),		
	[adult females]	37.3	Batieha et al. [144]
	[adult males]	5.1	
Lebanon	Prevalence (%),		
	[girls]	32	El-Hajj Fuleihan et al. [145, 146]
	[boys]	9–12	
Morocco (Rabat)	Prevalence (%), [women]	91	Arabi et al. [147]
Saudi Arabia	Prevalence (%), [girls], [adolescent]	Up to 80	Siddiqui and Kamfar [148]
Tunisia (Ariana)	Prevalence (%), [women], [Age: 20–60]	47.6	Arabi et al. [147]
Turkey (Ankara)	Prevalence (%),		
	[mothers]	46	Arabi et al. [147]
	[newborns]	80	
Iron deficiency			
Arab Gulf countries	Prevalence (%),		
	[children], [preschool age]	20–67	Musaiger [149]
	[children], [school age]	12.6–50	
	[pregnant women]	22.7–54	
Bahrain	Prevalence (%),		
	[children], [Age: 6–59 months]	48	Bagchi [150]
	[women], [Age: 15–49]	37.3	
Egypt	Prevalence (%),		
	[children], [Age: 6–59 months]	25	Bagchi [150]
	[women], [Age: 15–49]	11	
Iran	Prevalence (%),		
	[children], [Age: 6–59 months]	15–30	Bagchi [150]
	[women], [Age: 15–49]	33.4	
Jordan	Prevalence (%),[children], [school age]	20	Bagchi [150]
	Prevalence (%), [women], [Age: 15–49]	28	

TABLE 2: *Cont.*

Lebanon	Prevalence (%), [children], [Age: 6–59 months]	23	Bagchi [150]
Morocco	Prevalence (%),		
	[children], [Age: 6–59 months]	35	Bagchi [150]
	[women], [Age: 15–49]	30.1	
Oman	Prevalence (%),		
	[children], [Age: 5–14]	41	Bagchi [150]
	[women], [Age: 15–49]	40	
Pakistan	Prevalence (%),		
	[children], [Age: 6–59 months]	60	Bagchi [150]
	[women], [Age: 15–49]	30	
Palestine	Prevalence (%),		
	[children], [Age: 6–59 months]	53	Bagchi [150]
	[women], [Age: 15–49]	36.2	
Saudi Arabia	Prevalence (%), [children], [preschool age]	17	Bagchi [150]
Syria	Prevalence (%),		
	[children], [Age: 6–59 months]	23	Bagchi [150]
	[women], [Age: 15–49]	40.8	
UAE	Prevalence (%),		
	[children], [Age: 6–59 months]	34	Bagchi [150]
	[pregnant women]	14	
Yemen	Prevalence (%), [children], [preschool age]	70	Bagchi [150]

Different limits of blood levels define VDD, ranging from insufficiency to severe deficiency, similar for Iron deficiency. UAE: United Arab Emirates [147, 150].

In the other category of diseases, deficiencies in vitamins and minerals affect particularly children and women of childbearing age. MNDs impair physical and mental development of children, exacerbate infections and chronic diseases, and impact morbidity and mortality. Most countries of the MENA have widespread MNDs, yet countries of the Gulf Cooperation Council, Iran, and Tunisia have moderate levels of MNDs [20]. Among MNDs, iron deficiency is the most prevalent nutritional problem in the MENA [21]. Its prevalence in the region varies from 17 to 70% among preschoolers, 14 to 42% among adolescents, and 11 to more than 40% among women of childbearing age. Severe iron deficiency is a direct cause of anemia. Nonnutritional genetic anemias are known to be relatively common in the region but will not be tackled because they are not affected by dietary factors. Nutritional deficiencies other than iron, such as folic acid, vitamin B12, and vitamin C, are also prevalent in some countries of the region but data are scarce [20]. The Middle Eastern and South Asian regions have the highest rates of Vitamin D Deficiency (VDD) worldwide [22]. The prevalence ranges of VDD in the MENA are 46–83% for adolescents and adults, and 50–62% among veiled women [5]. It reaches up to 70% in Iran [23] and 80% in Saudi Arabia [24]. Table 2 shows prevalence of VDD and iron deficiency in countries of the MENA where data are available.

2.3 THE DIETARY SIGNATURE ON THE GENOME

The effect of food on gene function is the focus of nutritional genomics, an emerging field of study that focuses on the molecular, cellular, and systemic levels of this effect [25, 26]. The abundance of calories, macro- and micronutrients, and bioactive food elements constitutes the nutritional environment that alters the genome, the epigenome, the posttranscriptional regulation, and the posttranslational modifications, leading to a variety of metabolic functional gene-products, as shown in Figure 5. The nutritional environment channels the function of gene-products into certain pathways, preferring certain biological activities over others and resulting in the final phenotypic outcome and health-disease status. Correspondingly, two disciplines of nutritional genomics are entertained: nutrigenomics and nutrigenetics. The former started as a focus on the effect of nutrients on gene expression, while the latter, yet a different field of study, emerged in search of approaches to alter the clinical manifestations of certain rare diseases, such as certain inborn errors of metabolism, via personalized diet. The link between them is the functional gene product. It is the end-point in nutrigenomics and the starting point in nutrigenetics. Other conventional terminologies relevant to the diet-genetics-disease relation will also be treated in the following section, prior to discussion of the dietary signature on the genome in the MENA.

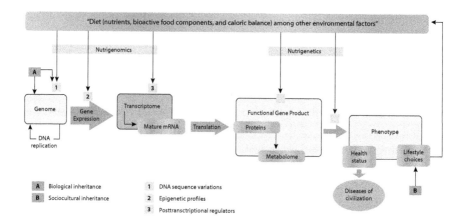

FIGURE 5: The dietary signature. The biologically inherited DNA genome accumulates DNA sequence variations over generations. Epigenetic profiles determine which parts of it are to be transcribed. Once transcribed to RNA, it matures into different mature RNA outcomes depending on the post-transcriptional regulators. Among the effects of the environment on DNA sequence variations, epigenetic profiles, and post-transcriptional regulators, the effect of diet is studied in nutrigenomics. After translation, and under the impact of dietary status surrounding the primarily translated proteome, the final set of functional proteins, activated pathways, and subsequent metabolites constitutes the functional Gene Product. The gene product is only potentially functional towards a certain phenotypic outcome. The downstream end result of health status depends greatly on what nutrients are fed into the systemic machine of gene products. The functional gene product is the end-point in nutrigenomics and the starting point in nutrigenetics. It is a marker of the phenotypic outcome: expression of disease and prognosis. Phenotype may dictate the lifestyle choices available to a certain individual, including taste preferences, which are also delineated by culturally inherited customs and habits. In their turn, lifestyle choices including dietary habits determine environmental exposures. Furthermore, civilization diseases have been hidden for a long period of time due to the sociocultural inheritance of adequately evolved matching lifestyle preferences and diet choices that have been masking a biologically inherited limited gene pool. The genes being in status quo, in presence of a nutritional transition, the rates of civilization diseases are on the rise because of the loss of the protective adequacy of the diet. This highlights the presence of hidden genes, the phenotypic expression of which can be masked by a specific nutritional state, such as that corresponding to the Mediterranean diet, as more increasingly being recommended recently in the literature. However this cannot be answered if sequence variations and specific SNPs affecting nutritional needs are not tested for in the specific populations.

1. The *genome sequence* of DNA base pairs dictates the primary ge-
    netic profile and hence gene function. DNA sequence variants—
    gene variants and single nucleotide polymorphisms (SNPs)—des-
    ignate an alteration of gene structure with or without functional
    changes that might or might not lead to a complex gene-function
    relationship depicted in different diseases [27]. The epigenome, an
    emerging concept in genetic research, is a set of nongenetic factors,
    such as diet, that change the expressed gene outcome without af-
    fecting the structure of the DNA per se.

2. *Nutrigenomics* considers environmental factors of alimentary
    source that may disrupt the DNA sequence in peptide-coding
    and in promoter regions, affecting the gene product. Other envi-
    ronmental nutrigenomic factors include abundance of macro and
    micronutrient components of the diet, presence of other bioactive
    food elements, and caloric content. Under- or overnutrition in the
    maternal environment sets epigenetic programming mechanisms
    via energetic control of function and oxidation. Through regula-
    tion of many biological functions including mitochondrial activity,
    cellular stress, inflammation, and telomere shortening, the dietary
    signature starts when epigenetic mechanisms induce or limit the
    risk to disease [28]. Possible levels of expression of a certain gene
    lie in a range of disease susceptibility that is determined by epigen-
    etic mechanisms. These mechanisms are dictated by the functional
    profile of the cell, which obeys its nutritional state and reflects the
    nutritional environment.

3. Although sustainability of *epigenetic programming* along life span
    is not well understood [28], two temporally distinct profiles may be
    distinguished. First, the basal epigenome is determined early on in
    life. Depending on the basal expressivity of DNA, it behaves like
    a permanently edited version of the genome. Accordingly, increas-
    ing evidence of trans-generational inheritance of epigenetics was
    found in mice [29] through the effect of grandmaternal nutrition on
    grandchildren during gamete stage, throughout the mother's fetal
    stage [30]. Second, later in life, similar mechanisms affect gene ex-
    pressivity in response to temporary environmental factors, resulting
    in a short-lived epigenetic profile. These changes are mainly due

to interference of nutrients and bioactive food components with transcription factor conformations [26]. This signature serves as a means for the organism to receive information about its nutritional environment in order for the cells to execute appropriate modifications on the profile of expressed genes [31]. The nutrigenomic signature is not well studied in humans yet; however obvious importance is due to its impact on gene expression, chronic diseases susceptibility, and health status of future generations [31–33].

4. Regulators of *posttranscriptional modifications* affect alternative RNA splicing which gives rise to different mature mRNA isoforms. Alternative splicing is as highly prevalent as in 35 to 59% of human genes [34]. Post-trancriptional regulators, such as microRNAs and their coacting and counteracting proteins, are part of the RNA and protein pools [35]. They are hence influenced by epigenetic and metabolic factors as well [26, 36].

5. *Proteomics* is the study of the protein pool in the organism, as an integral part of the cellular function. On the other hand, the metabolome designates the structure, the localization, the post-translational modifications, and the functions of proteins and metabolites along with their interactions in the organism [37]. It defines the current metabolic state and active intracellular pathways in the organism (Figure 5). The functional gene products comprise all the potentially functional molecules and pathways, whether currently active or not, that result from a certain genome-epigenome combination, leading to a certain range of possible phenotypic outcomes, rather than a clearly defined health status.

6. *Nutrigenetics* is a quite different approach that emerged when dietary interventions were able to successfully alter the course of certain diseases. The basic principle considers how the same dietary environment can result in different phenotypic outcomes of health or disease in metabolizers with different functional gene-products or programmed phenotypes [38]. The concept is similar to how individuals possess different phenotypes as drug-metabolizers. The study of genetic variations affecting nutrient metabolism, from digestion to detoxification, can decipher ambiguities in the diet-disease relationship [38]. However, the challenge lies in the ability

of researchers to describe the processes through which the dietary environment imposes itself to precipitate metabolic disorders.

7. Finally, hypotheses of *Thrifty Profile*, namely thrifty genes and thrifty phenotypes, offer explanations for etiology, predisposition, and rising prevalence of DM and obesity. Early life dietary habits foretell the basal appetite control and cellular nutritional needs through psychological and molecular habituations [28]. Thrifty genes that enable survival during periods of food shortages may have been conserved over generations under the selection pressure of under-nutrition [39]. Thrifty phenotypes may be due to early nutritional challenges that enhance nutrients-saving mechanisms in the growing individual, leading to excessive storage later on and increased risk of metabolic disorders [40]. Both models have not gathered enough evidence apiece; however combined they provide a fertile base for further nutritional genomic research.

An example of a phenotype that has evolved accordingly is taste preferences and ability to digest, absorb, and appropriately respond to nutrients [41]. Genes for taste receptors, among other proteins that handle the metabolism of different nutrients, have been extensively studied. In an extensive review by Garcia-Bailo et al., an important aspect of the dietary signature is addressed: the genetic variations that affect dietary habits and food choices, with an emphasis on their effects on the nutritional environment and the health outcome [41].

Given the rise in multifactorial diseases, nutrigenetics started to involve public health research, hinting at personalized dietary recommendations for prevention of civilization diseases many years before clinical manifestations arise [31]. Adequacy of the general dietary recommendations to the ancient nature of our genes is becoming increasingly dubious. The human genome, as we know it, was sculpted throughout 2 million years of evolution under the diets of our hunters-gatherers ancestors [42]. Later on, the available food choices changed since the introduction of agriculture, but too rapidly for the ancestral stone-age genome to keep up with. This fast nutritional transition revealed evolutionary origins of obesity and diabetes among other civilization epidemics [43]. The experiment-based advancement of dietary recommendations during the past 25 years showed a convergence towards what looks more like a Paleolithic hunter-gatherer diet [44]. Despite low compliance to recommended diets and increasing industrialization of actual dietary habits, personalized and ancestral dietary recommendations still seem promising.

2.4 NUTRITIONAL GENOMICS IN THE MENA

Diet, genetics, and disease are linked in many ways as could be shown in Figure 5. The MENA is a region that has been witnessing simultaneously a dietary change and a worsening prevalence of chronic diseases. Because of this, nutritional genomics research in such a region can improve our understanding of this rapid change in disease prevalence and shed light on the genomic effects of this dietary transition in the region. Nutritional genomics research in the MENA is minimal. To our knowledge, this is the first review on the topic in the region. We aim to collate studies in MENA countries that discuss any aspect of the dietary signature that we discussed in Figure 5. We approach that using examples of common diseases with rising prevalence in the MENA. The paper discusses two categories of diseases: (1) civilization disorders of metabolism (cardiovascular diseases and metabolic risk factors), and (2) micronutrient deficiencies (MNDs).

We also look at other populations where nutritional genomics research in these disease categories was done and discuss how it applies to our region with recommendations for future research on MENA populations.

2.5 DIET-RELATED CIVILIZATION DISORDERS OF METABOLISM IN THE MENA REGION

Populations of the MENA belong to a unique genetic pool because of the mixture of ethnicities with horizontal mixing of populations throughout history, the high rate of consanguineous marriages within subpopulations, and the geography of the states making up the region (Figure 1). Nevertheless, wide prospective population studies on the effects of polymorphisms on such disorders of metabolism in the MENA are still lacking [25]. Based on literature reports on other populations, a large set of genes and DNA sequence variants are potentially culpable of the rise of metabolic disorders under the effect of industrialized diets. In this part of the paper, we present numerous polymorphisms that predispose to metabolic disorders including T2DM, obesity, dyslipidemias, atherosclerosis, cardiovascular events, and hypertension. The civilization disorders have common pathophysiologies and risk factors of metabolism and, since interrelated and comorbid, will accordingly be treated as one major health outcome in the following discussion about genetic entities common to the different disorders, under the effect of diet.

The *Brain-Derived Neurotrophic Factor* (*BDNF*) is important for energy balance in mice and for regulation of stress response in humans (OMIM 113505). Polymorphisms in this gene are associated with obesity and all subtypes of psychological eating disorders in Europeans (NCBI 627). Recently, three-way association was identified between hoarding behavior of obsessive-compulsive disorder, obesity, and the Val/Val genotype of *BDNF* in the Valine (Val) to Methionine (Met) amino acid change at position 66 (Val66Met) in Caucasians [45]. The suggested evolutionary mechanism for this complex relationship between gene, psychopathology, and body weight is the conservation of a thrifty gene, once an old survival strategy.

Control of fetal appetite was recently shown to be a function of the *Fat Mass- and Obesity-Associated* (*FTO*) gene expression [28] which codes a nuclear oxygenase that affects tissue lipid metabolism [46] (NCBI 79068) and depends on energy balance during development [28]. A strong relation links *FTO* SNPs to higher risk of obesity and T2DM in many international studies (OMIM 610966) [47, 48].

Transcription Factor 7-Like 2 (*TCF7L2*) gene codes a transcription factor involved in blood glucose homeostasis. The rs7903146 variant association to T2DM varies greatly over ethnicities (OMIM 602228). However in Palestinians, this SNP (114758349C > T) increases the risk for T2DM, and homozygotes are affected at younger age [49].

Calpain 10 (*CAPN10*), which codes a calcium-dependent cysteine protease, is being increasingly studied for its role in T2DM (NCBI 11132). SNP-44 of *CAPN10* has significant association with T2DM and total cholesterol in Gaza [50], while only UCSNP-19 SNP and haplotype-111 are proven to be high risks for T2DM in Tunisia [51].

The association of T2DM with polymorphisms of the *Angiotensin Converting Enzyme* (*ACE*) and the *Methylene Tetrahydrofolate Reductase* (*MTHFR*) is not well proven [52]. However data in Tunisians suggest synergistic action of the *ACE* Insertion/Deletion (I/D) dimorphism with the *MTHFR* C677T SNP on risk of T2DM [52]. Fairly common, *ACE* D and *MTHFR* 677T alleles are, respectively, present in around 77 and 27% of Moroccans [53]. Nevertheless, *ACE* DD genotype in Tunisians is associated with higher ACE activity and might become a useful clinical marker for CAD risk assessment of acute myocardial infarction [52, 54]. In Lebanon, *ACE* D allele and age, combined, are associated with higher risk for hypertension [54]. Also, Lebanese with *MTHFR* C677T turned out to be more susceptible to diabetic nephropathy than Bahrainis with the same SNP [55]. The SNP cannot constitute an independent risk factor in Arabs [56]. Its effect is presumably due to high homocysteine levels and hence must be evaluated depending on dietary and ethnic backgrounds [55].

In genes encoding the G protein-coupled Beta-2- and 3-Adrenergic Receptors (*ADRB2, ADRB3*), evolutionary selection of specific alleles exists in Africans, Asians, and Europeans. *ADRB2* Glu27 and Gln27 are, respectively, factors of exercise-dependent obesity risk and metabolic syndrome susceptibility (OMIM 109690). Glu/Glu and Glu/Gln can independently predict severe Coronary Artery Disease (CAD) in Saudi Arabs [57]. However *ADRB3* Trp64Arg SNP is a CAD predictor only in presence of other risk factors in Arabs, but not an independent one [57]. ADRB3 is mainly located in adipose tissues causing easier weight gain and earlier T2DM onset in Trp64Arg individuals in several populations [58] (NCBI 155) (OMIM 109691).

Peroxisome Proliferator-Activated Receptor Gamma (*PPARG*) genes encode nuclear receptors and regulators of adipocyte differentiation and possibly lipid metabolism and insulin sensitivity (OMIM 601487). Pro-12Ala isoform of *PPARG2* seems to activate transcription less effectively and carry less morbidity. Carriers of a Pro12Ala polymorphism may have a weaker BMI correlation to amount of dietary fat when compared to Pro homozygotes [59], while response to quality of dietary fat is greater in terms of BMI, lipid profile, and fasting insulin levels [60, 61]. However these associations were not found for many of the studied populations (OMIM 601487).

Apolipoproteins (APOs) are involved in lipid metabolism. *APOE* polymorphisms have been heavily studied. In the *APOE* G219T SNP, TT individuals have prolonged postprandial lipemia [62]. Apo E has three major isoforms, E2, E3, or E4. APOE E4 individuals may be protected effectively by lower dietary fat intake [63] while non-E4 individuals have minimal to no benefit from dietary intervention on lipid profile [64]. In Iranians, E2 allele was associated with lower total cholesterol levels [65]. However, despite correlation between APOE2 and LDL subfraction profiles in healthy Arabs, no similar association was found in Arabs with CAD [66]. *APOE* E2, E3, and E4 carriers constitute approximately 11, 79, and 10% of Moroccans, respectively [53].

Moreover, mutations in *Lipoprotein Lipase* (*LPL*), which is crucial for receptor-mediated lipoprotein uptake, drastically affect lipoprotein metabolism disorders (NCBI 4023). In Saudi Arab population however, lack of association between *LPL* polymorphisms and CAD was noticed [56]. Strong evidence exists for Hepatic Lipase (*LIPC*) C514T homozygotes. They have more atherogenic lipid profile in response to dietary fat in addition to impaired adaptation to higher animal fat with higher cardiovascular diseases risk [67, 68].

Finally, the Paraoxonase (*PON1*) gene encodes for an anti-atherosclerotic esterase which capacitates high-density lipoproteins to prevent lipoprotein oxidation. Gln192Arg and Leu55Met are two common polymorphisms of *PON1* that modulate PON1 activity in the serum, which predicts the architecture of apolipoprotein, lipoprotein, and lipid levels [69]. In the late 1990s, PON1 status, including genotype and serum activity levels, has been proven to predict cardiovascular risk much better than genotype alone [70]. However, in Turkish subjects, there was no consistent association between the polymorphisms and the lipid levels [71]. An individual's polymorphism might be suggestive of a high risk while his dietary signature is making the actual PON1 activity favorable, that is, low risk. This interaction between diet and genes can hinder the significance of genetic screening, and enhance the relevance of proteomics and metabolomics. The lesson learned from the *PON1* role in cardiovascular disease is of utmost relevance. Functional genomic analysis is required for adequate risk assessment; an individual may be screened for all known polymorphisms of *PON1*, but still not be assigned a risk category for cardiovascular disease [72].

Discrepancies between genotype and function impose limitations on genetic screening. Similarly for most of the polymorphisms presented previously, the degree to which genetic screening can be helpful in decision-making is controversial. More activity correlation studies are needed to examine the "penetrance" of polymorphisms. Also, insufficient nutrigenomic and proteomic evidence may be misleading [117]. Hence, further multidisciplinary studies, with coordination between laboratories, will be needed to decide which gene/polymorphism would be worth screening in a particular population.

Further multidisciplinary nutritional genomics research is needed for more specific targeted individualized advising and therapy. However, given the current lack of adequate understanding of the genetic etiologies of civilization diseases and wide-scale regional genetic screening studies, reversal of dietary changes is rendered the simplest available measure to control the metabolic epidemic of civilization diseases in the MENA.

2.6 MICRONUTRIENT DEFICIENCIES (MNDS) IN THE MENA REGION

Micronutrients (vitamins and minerals) are required throughout life, in minute amounts in the human body, to function as cofactors of enzymes or as structural components of proteins, or to maintain genome stability, among other physiological roles [118]. Both their excess and deficiency may cause DNA damage, alter growth and development, contribute to a wide array of chronic diseases, and jeopardize health [118]. MNDs are highly prevalent in MENA countries as was established earlier. Deficiencies in iodine, iron, and vitamin A are very important MNDs in terms of prevalence and potential threat to public health worldwide; however relevant gene-diet interaction has not been sufficiently studied in the MENA. This section will thus be restricted to the following MNDs of particular interest in diet-genetics-disease interaction: vitamin D, calcium, iron, folate, and vitamins C, E, B6, and B12.

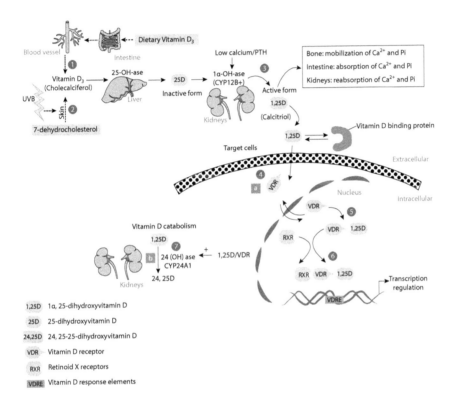

FIGURE 6: Vitamin D pathway and sites of interaction with dietary factors. Cutaneous or dietary vitamin D is hydroxylated in the liver to form 25-hydroxyvitamin D (1,2) and in the kidney to form 1α,25-dihydroxyvitamin D (3). 1α,25-dihydroxyvitamin D binds to VDR (4), the 1α,25-dihydroxyvitamin D ligand promotes VDR-RXR heterodimerization (5), and the complex binds to VDRE to mediate transcriptional regulation of target genes (6). The concept of gene-diet interaction is described in the vitamin D pathway by the different polymorphisms in the VDR gene (a) and the dietary regulation of CYP24A1 enzyme (b).

2.6.1 VITAMIN D DEFICIENCY (VDD)

Vitamin D is a fat-soluble vitamin, with two forms, one present in a narrow range of foods (D2) and another formed under the skin when exposed to the ultraviolet B (UVB) light fraction of sunlight (D3); both are activated by the liver and kidneys [119] (Figure 6). Prolonged VDD can result in rickets in young children and osteoporosis and fractures in adults [120]. Recently, low vitamin D levels have been associated with increased risk of hypertension, cardiovascular diseases [121], cancer [122], diabetes [123], musculoskeletal and immunity disorders, and infectious diseases [124].

Despite the sunny climate, the MENA has a highly prevalent VDD across all age groups, with the highest rate of rickets worldwide [22]. The main reasons are limited sun exposure and low dietary vitamin D intake, along with frequent pregnancies, short breastfeeding periods [125], skin pigmentation [126], body mass index [127], religious practices [128], and educational levels [129].

In addition to nutritional and social factors of VDD, genetic factors also play an important role and are depicted on the metabolic pathway of vitamin D shown in Figure 6. Genetic variations predisposing to VDD are related to Vitamin D Receptor (*VDR*) polymorphisms at intron 8 (BsmI) and exon 2 (FokI) [130]. The Fok1 polymorphism (C>T) in the translation initiation site creates an upstream initiation codon and a three amino acids longer molecule in the f allele compared to the F allele [131] which gives a more transcriptionally active VDR [132] leading to the tolerance to low vitamin D levels observed in Egyptian FF homozygotes [130]. This suggests possible evolutionary adaption to dietary intake or lifestyle changes. Only FF homozygote children have increased calcium absorption and bone mineral density [131]. Conversely, the decreased calcium absorption linked to the f allele was correlated with an increase in colon cancer risk only when calcium dietary intake is low [133]. FF genotype seems hence more advantageous than ff genotype. Paradoxically, FF (shorter VDR) was correlated with rickets unlike ff (longer VDR) in Turks and Egyptians. Thus, further studies are needed to understand the complex genetics and risks of rickets. *VDR* B allele also predisposes to VDD since Egyptian B homozygotes had severe rickets [130]. In other studies in the Middle East, high vitamin D doses were needed to treat patients with rickets [134].

Thus, the unexpected high prevalence of VDD in the MENA could be linked to VDR polymorphisms.

Moreover, a G>A polymorphism at position −3731 of the cdx-2 (Caudal-Type Homeobox Transcription Factor) binding element on the *VDR* gene promoter is another genetic variant of *VDR* that affects calcium absorption. The Cdx2 promoter A allele (cdx-A) binds cdx2 more strongly and has a greater transcriptional activity compared to the cdx-G allele. Thus, the A allele may increase intestinal *VDR* expression, subsequently enhancing calcium intestinal absorption and increasing bone mineral density. The differential expression of *VDR* shows how genetic differences influence the body response to nutrients [135].

Bioactive food components may exert an effect on gene expression and enzyme activity, subsequently decreasing disease risk. For example, 4′,5,7-Trihydroxyisoflavone (genistein), a soy component, is a genome-protective nutrient. It inhibits the activity of CYP24A1 (Figure 6) and thus 1,25$(OH)_2$D degradation, increasing VDR stability and the half-life and biological effects of vitamin D [136]. Folate can also inhibit this activity by increasing methylation of the promoter of *CYP24A1*. Also, addition of vitamin D and calcium to the western diet significantly decreases the incidence of colon cancer [133].

2.6.2 IRON DEFICIENCY

Iron (Fe) is an essential mineral needed in small amounts mainly for the production of hemoglobin and utilization of oxygen among other vital functions. Iron deficiency is a common MND mostly caused by low intake of iron, blood loss, and parasitic infections [137]. Iron absorption is enhanced by vitamin C, low pH, and heme iron and hindered by bioactive vegetables components (polyphones, tannins, phytates) and calcium [138].

Genetic factors that contribute to iron deficiency, in addition to the dietary intake, are underscored by the ability of many individuals to maintain normal iron levels despite low iron dietary intake. Nutritional iron deficiency and genetic iron deficiency have been experimentally distinguished in mice. Hephaestin (Heph) is a multicopper oxidase that allows iron basolateral surface export. Sex-linked anemia (*Sla*) mice bearing a deletion in *Heph* gene compared to control mice showed different responses to diet; *Sla* mice had duodenal iron accumulation and low plasma iron [139]. Additionally, mutations in the genes of human hemochromatosis protein (*HFE*) and its interacting protein beta-2 microglobulin (*B2M*), which play an important role in iron metabolism, cause murine iron deficiency [140].

2.6.3 FOLATE DEFICIENCY

Folic acid is a water soluble B vitamin of exclusive dietary origin. It provides the one-carbon metabolism with its main coenzyme form, tetrahydrofolate (THF). A key enzyme herein, methylenetetrahydrofolate reductase (MTHFR), catalyzes vitamin B12-dependent conversion of homocysteine to methionine, a precursor of S-adenosylmethionine (SAM), a methyl donor to DNA [141]. Thus, folate deficiency results in hyperhomocysteinemia (HHC), a risk factor for CAD [142].

MTHFR C677T is prevalent in 4% of Pakistanis [32], 2% of Yemenite Jews, 10% of Muslim Arab Israelis, and 11% of Lebanese [151]. It leads to a thermolabile MTHFR, which precipitates HHC in low folate states. This gene-environment combination is a risk factor for cardiovascular diseases [152], neural tube defects, and other chronic diseases [153].

C677T is a genetic variation that affects individual dietary requirements because it makes prevention of folate-related diseases that require higher folate intake. Korean C homozygotes develop HHC only with low folate levels, while T homozygotes have HHC even with normal folate levels [154]. Surprisingly however, the hyperhomocysteinemia, low folate levels, and increased cardiovascular risk observed in Indian Asians compared to European whites were not attributed to the *MTHFR* 677T variant [155]. Studies specific to the MENA are therefore needed because of possible variations from other regions.

The example of MTHFR and folate also underscores the genome-epigenome interplay. The TT genotype alters DNA methylation and gene expression in peripheral blood mononuclear cells only in folate deficient patients [156]. Paradoxically, the same polymorphism is inversely associated with and hence has a protective role against colorectal cancer (CRC). CRC risk decreases with adequate methionine intake, which leads to an increased formation of SAM and a negative feedback inhibition of MTHFR activity. However, the protective roles of MTHFR mutation and methionine dietary intake require an adequate dietary folate intake [157]. These findings suggest that inadequate folate intake puts carriers of particular genetic variants at higher risk of cancer. Thus personalized dietary interventions might be beneficial in reducing cancer risks.

Serum vitamin B12 deficiency is highly prevalent in Iran [158]. Thus, low folate and vitamin B12 levels can be inversely linked to the hyperhomocysteinemia observed in this population. Importantly, high prevalence of CAD was observed in a Turkish population with low plasma folate despite low plasma cholesterol concentrations [159]. It remains to determine the link between hyperhomocysteinemia, low folate intake, CAD risk, and MTHFR variants, to explain this prevalence in Turkey and possibly in other countries of the MENA region. This would add folate supplementation as a possible treatment for hyperhomocysteinemia. Other studies have also reported the possible contribution of the level of pyridoxal phosphate (PLP or vitamin B6) to hyperhomocysteinemia and vascular disease; low PLP levels were observed in individuals with the homozygote TT genotype compared to healthy individuals from some countries of the MENA [160, 161]. These data corroborate the various dietary signatures on the specific genetic profile and show that other mechanisms, enzymes, and vitamins should be examined as well.

2.6.4 VITAMIN B12 DEFICIENCY

Like folate deficiency, vitamin B12 (cobalamin) deficiency can affect establishment of the disease depending on the genetic background. A common genetic variant is detected in Methionine Synthase Reductase (MTRR), an important enzyme for maintaining Methionine Synthase in its active state. The polymorphism is an A66G substitution resulting in an Ile22Met residue. The homozygous genotype was associated with an increased risk of neural tube defects (NTDs) when combined with low vitamin B12 levels. Vitamin B12 deficiency is highly prevalent among Iranian women of childbearing age [158]. However polymorphisms in both MTHFR and MTRR increase NTDs risk [162].

2.6.5 VITAMINS C AND E DEFICIENCIES

As antioxidants, vitamins C and E have an important function in the diet-gene interaction. Glutathione S-transferases (GSTs) transfer glutathione to different substrates. A common deletion of *GSTM1* gene, a deletion polymorphism in *GSTT1*, and an A313G polymorphism of *GSTP1* result, respectively, in a nonfunctional genotype, loss of enzyme activity, and altered activity of the *GST* isoforms. The GST enzymes were found protective against serum ascorbic acid deficiency when vitamin C consumption is low, since GST null genotypes with low vitamin C intake had an increased serum ascorbic acid deficiency risk [163]. The Hp1 and Hp2 polymorphisms in the hemoglobin-binding protein haptoglobin (Hp) were also studied in vitamin C deficiency. Unlike Hp1 carriers, Hp2 homozygotes had lowest serum vitamin C concentrations when dietary vitamin C intake is insufficient. Thus, Hp1 has a greater antioxidant capacity preventing hemoglobin-iron-related vitamin C oxidation and depletion [164].

A protective role for vitamin E against atherosclerosis, cancer, and neurodegenerative diseases has also been reported. Polymorphisms in the proteins involved in vitamin E metabolism lead to differential vitamin E uptake and response among individuals, and subsequently different disease risk [165]. Dietary vitamin E intake also influences the body mass index (BMI) and risk of obesity via modifying genetic variants of SIRT1 (sirtuin protein family of nicotinamide-adenine-dinucleotide- (NAD+)-dependent histone deacetylases) [166].

Well-studied gene-diet interactions are also critical in the pathophysiology of cancer. Given the evidence that VDR polymorphisms, MTHFR genotype, and DNA methylation in a low calcium or folate intake are associated with an increased cancer risk, the dietary signature greatly influences carcinogenesis. Significant dietary factors include antioxidants such as vitamin C, carotenoids, lycopene, tocopherols (vitamin E), and many other micronutrients present in fruits and vegetables.

Being one of the leading causes of death worldwide as well as in the MENA, cancer has been extensively studied, and the diet-genetics-cancer interaction is currently being thoroughly investigated for each and every one of the involved micronutrients. So, the effect of the diet-genetics interaction on carcinogenesis will not be dwelled upon in this paper.

The aforementioned studies collectively depict the interaction between diet, genetic variability, and disease. Genetic variants might affect gene expression patterns and epigenetic events resulting in differential body responses to diet. However, a small individualized nutritional intervention that is well studied to provide the needed concentrations of micronutrients can influence genetic variants to decrease disease risk. Thus, nutrigenetics is a tool for choosing the appropriate diet according to the individual's genetic makeup.

2.7 A CALL FOR NUTRITIONAL GENOMICS RESEARCH IN THE MENA

All previous data support the crosstalk between diet and genome. Differential responses to dietary components among individuals are determined by genetic factors. The deleterious effects of some genotypes can be circumvented by an increased intake of particular nutrients to overcome the genetic susceptibility, which opens the horizon for personalized diet. In turn, nutrients might affect genome, gene expression, and phenotype. Although hard and complex, it is worthwhile to identify the genes that predispose individuals to chronic diseases and the nutrients that regulate their expressions to modify personal risks and to prevent, mitigate, or treat diseases. Studying on an individual basis the interactions between diet and genetics could help select appropriate diet to optimize health status. Indeed, the picture becomes more complicated when lifestyle, behavioral, and other environmental factors interfere with the diet-genetics interaction.

Of note is the unique ethnic combination of the region's native populations that make studies from other regions inapplicable to the MENA. In spite of the multiethnic origins, high rates of consanguinity in the subpopulations render the genetic pool paradoxically limited and significantly increase not only the risk of congenital abnormalities but also the susceptibility of the population to chronic diseases and genetic disorders [167]. Screening for the common polymorphisms in the MENA can give insights on their prevalence in the region or can help discover polymorphisms indigenous for the region. Such action could help alleviate the burden of chronic diseases in the MENA simply by suggesting adequate adjustments of dietary factors to hide a genetic polymorphism or to prevent DNA damage.

In some of the examples provided previously, success can be achieved, but in others, researchers ought to be more cautious. Selecting which polymorphisms are to be screened for, at the population level in the MENA, should be made after careful understanding of the effects of these variants on diet and disease. This is crucial to avoid misleading results and unnecessary costs. Unfortunately, functional studies are limited, but wide-scale screening and associations from other regions in the world could guide the decision-making process regarding screening in the MENA. At the same time, more effort and money should be invested in molecular and cellular research in nutritional genomics in order to better understand the function of the dietary signature and to more confidently guide population screening and personalized diet.

2.8 CONCLUSIONS

1. MENA countries are witnessing a radical change in dietary patterns from a traditional diet to a less healthy industrialized diet.
2. Rising prevalence for civilization diseases of metabolism and micronutrient deficiencies in the MENA parallels the change in dietary habits and is mostly caused by it.
3. Nutrigenomic factors and the dietary signature on the genome play a role in the diet-disease interactions.
4. Genetic sequence variations, epigenetic profiles, and posttranscriptional and posttranslational modifications are some of the mechanisms that define the diet-genetics-disease relationship.
5. A large set of gene polymorphisms have been correlated with civilization diseases of metabolism, only a little of which have been studied in MENA countries.
6. There are different mechanisms through which diet-genetics interaction affects micronutrient pathways and contributes to disease, including vitamin D and calcium, iron, folate, and vitamins C and E.

7. Given the drastic dietary changes in the region over a short period of time, diet is the most obvious public health intervention, yet system biology and genomics research should not be underestimated.
8. Wide-scale screening for certain gene polymorphisms in the MENA might allow for efficient intervention with personalized diet.
9. More nutrigenomics research is needed to look at function and mechanisms of the diet-genetics-disease interaction.

REFERENCES

1. D. Grigg, "Food consumption in the Mediterranean region," Tijdschrift voor Economische en Sociale Geografie, vol. 90, no. 4, pp. 391–409, 1999.
2. A. O. Musaiger, "Diet and prevention of coronary heart disease in the Arab Middle East countries," Medical Principles and Practice, vol. 11, supplement 2, pp. 9–16, 2002.
3. A. Belal, "Nutrition-related chronic diseases Epidemic in UAE: can we stand to STOP it?" Sudanese Journal of Public Health, vol. 4, pp. 383–392, 2009.
4. A. Alwan, "Noncommunicable diseases: a major challenge to public health in the Region," Eastern Mediterranean Health Journal, vol. 3, pp. 6–16, 1997.
5. K. Bagchi, "Iron deficiency anaemia—an old enemy," Eastern Mediterranean Health Journal, vol. 10, no. 6, pp. 754–760, 2004.
6. D. C. Klonoff, "The beneficial effects of a Paleolithic diet on type 2 diabetes and other risk factors for cardiovascular disease," Journal of Diabetes Science and Technology, vol. 3, no. 6, pp. 1229–1232, 2009.
7. A. Trichopoulou, T. Costacou, C. Bamia, and D. Trichopoulos, "Adherence to a Mediterranean diet and survival in a Greek population," The New England Journal of Medicine, vol. 348, no. 26, pp. 2599–2608, 2003.
8. S. Russeau and M. Batal, The Healthy Kitchen, Recipes from Rural Lebanon, Ibsar, Nature Conservation Center for Sustainable Futures American University of Beirut, 2009.
9. P. Mirmiran, R. Sherafat-Kazemzadeh, S. Jalali-Farahani, and F. Azizi, "Childhood obesity in the Middle East: a review," Eastern Mediterranean Health Journal, vol. 16, no. 9, pp. 1009–1017, 2010.
10. S. Yusuf, S. Reddy, S. Ôunpuu, and S. Anand, "Global burden of cardiovascular diseases part II: variations in cardiovascular disease by specific ethnic groups and geographic regions and prevention strategies," Circulation, vol. 104, no. 23, pp. 2855–2864, 2001.
11. P. Zimmet, D. Canteloube, and B. Genelle, "The prevalence of diabetes mellitus and impaired glucose tolerance in Melanesians and part-Polynesians in rural New Caledonia and Ouvea (Loyalty Islands)," Diabetologia, vol. 23, no. 5, pp. 393–398, 1982.
12. International Diabetes Federation, IDF Diabetes Atlas, 1st edition, 2000, http://www.eatlas.idf.org/.

13. International Diabetes Federation, IDF Diabetes Atlas, 2nd edition, 2003, http://www.eatlas.idf.org/.
14. International Diabetes Federation, IDF Diabetes Atlas, 3rd edition, 2007, http://www.eatlas.idf.org/.
15. International Diabetes Federation, IDF Diabetes Atlas, 4th edition, 2010, http://www.eatlas.idf.org.
16. S. Wild, G. Roglic, A. Green, R. Sicree, and H. King, "Global prevalence of diabetes: estimates for the year 2000 and projections for 2030," Diabetes Care, vol. 27, no. 5, pp. 1047–1053, 2004.
17. WHS, World Health Statistics Annual, 2005, http://www.who.int/whosis/whostat.
18. WHS, World Health Statistics Annual, 2011, http://www.who.int/whosis/whostat.
19. S. Yusuf, S. Reddy, S. Ôunpuu, and S. Anand, "Global burden of cardiovascular diseases. Part I: general considerations, the epidemiologic transition, risk factors, and impact of urbanization," Circulation, vol. 104, no. 22, pp. 2746–2753, 2001.
20. WHO, Draft Nutrition Strategy and Plan of Action for Countries of the Eastern Mediterranean Region 2010–2019, 2009.
21. O. M. Galal, "Micronutrient deficiency conditions in the middle east region: an overview," Public Health Reviews, vol. 28, no. 1–4, pp. 1–12, 2000.
22. A. Mithal, D. A. Wahl, J. P. Bonjour et al., "Global vitamin D status and determinants of hypovitaminosis D," Osteoporosis International, vol. 20, no. 11, pp. 1807–1820, 2009.
23. M. Moussavi, R. Heidarpour, A. Aminorroaya, Z. Pournaghshband, and M. Amini, "Prevalence of vitamin D deficiency in Isfahani high school students in 2004," Hormone Research, vol. 64, no. 3, pp. 144–148, 2005.
24. A. M. Siddiqui and H. Z. Kamfar, "Prevalence of vitamin D deficiency rickets in adolescent school girls in Western region, Saudi Arabia," Saudi Medical Journal, vol. 28, no. 3, pp. 441–444, 2007.
25. J. M. Ordovas and V. Mooser, "Nutrigenomics and nutrigenetics," Current Opinion in Lipidology, vol. 15, no. 2, pp. 101–108, 2004.
26. G. P. A. Kauwell, "Emerging concepts in nutrigenomics: a preview of what is to come," Nutrition in Clinical Practice, vol. 20, no. 1, pp. 75–87, 2005.
27. R. G. Gosden and A. P. Feinberg, "Genetics and epigenetics—nature's pen-and-pencil set," The New England Journal of Medicine, vol. 356, no. 7, pp. 731–733, 2007.
28. S. Sebert, D. Sharkey, H. Budge, and M. E. Symonds, "The early programming of metabolic health: is epigenetic setting the missing link?" American Journal of Clinical Nutrition, vol. 94, no. 6, pp. 1953S–1958S, 2011.
29. R. A. Waterland and R. L. Jirtle, "Early nutrition, epigenetic changes at transposons and imprinted genes, and enhanced susceptibility to adult chronic diseases," Nutrition, vol. 20, no. 1, pp. 63–68, 2004.
30. S. M. Singh, B. Murphy, and R. L. O'Reilly, "Involvement of gene-diet/drug interaction in DNA methylation and its contribution to complex diseases: from cancer to schizophrenia," Clinical Genetics, vol. 64, no. 6, pp. 451–460, 2003.
31. R. Debusk, "The role of nutritional genomics in developing an optimal diet for humans," Nutrition in Clinical Practice, vol. 25, no. 6, pp. 627–633, 2010.
32. P. J. Stover, "Influence of human genetic variation on nutritional requirements," American Journal of Clinical Nutrition, vol. 83, no. 2, 2006.

33. R. L. Jirtle and M. K. Skinner, "Environmental epigenomics and disease susceptibility," Nature Reviews Genetics, vol. 8, no. 4, pp. 253–262, 2007.

34. B. Modrek and C. Lee, "A genomic view of alternative splicing," Nature Genetics, vol. 30, no. 1, pp. 13–19, 2002.

35. W. Filipowicz, S. N. Bhattacharyya, and N. Sonenberg, "Mechanisms of post-transcriptional regulation by microRNAs: are the answers in sight?" Nature Reviews Genetics, vol. 9, no. 2, pp. 102–114, 2008.

36. C. D. Davis and J. Milner, "Frontiers in nutrigenomics, proteomics, metabolomics and cancer prevention," Mutation Research, vol. 551, no. 1-2, pp. 51–64, 2004.

37. A. Pardanani, E. D. Wieben, T. C. Spelsberg, and A. Tefferi, "Primer on medical genomics part IV: expression proteomics," Mayo Clinic Proceedings, vol. 77, no. 11, pp. 1185–1196, 2002.

38. A. El-Sohemy, "Nutrigenetics," Forum of Nutrition, vol. 60, pp. 25–30, 2007.

39. J. V. Neel, "Diabetes mellitus: a "thrifty" genotype rendered detrimental by "progress"?" American Journal of Human Genetics, vol. 14, pp. 353–362, 1962.

40. R. S. Lindsay and P. H. Bennett, "Type 2 diabetes, the thrifty phenotype—an overview," British Medical Bulletin, vol. 60, pp. 21–32, 2001.

41. B. Garcia-Bailo, C. Toguri, K. M. Eny, and A. El-Sohemy, "Genetic variation in taste and its influence on food selection," OMICS, vol. 13, no. 1, pp. 69–80, 2009.

42. S. B. Eaton and S. B. Eaton, "Paleolithic vs. modern diets—selected pathophysiological implications," European Journal of Nutrition, vol. 39, no. 2, pp. 67–70, 2000.

43. A. M. Prentice, "Obesity in emerging nations: evolutionary origins and the impact of a rapid nutrition transition," Nestle Nutrition Workshop Series: Pediatric Program, vol. 63, pp. 47–57, 2009.

44. M. Konner and S. Boyd Eaton, "Paleolithic nutrition: twenty-five years later," Nutrition in Clinical Practice, vol. 25, no. 6, pp. 594–602, 2010.

45. K. R. Timpano, N. B. Schmidt, M. G. Wheaton, J. R. Wendland, and D. L. Murphy, "Consideration of the BDNF gene in relation to two phenotypes: hoarding and obesity," Journal of Abnormal Psychology, vol. 120, no. 3, pp. 700–707, 2011.

46. C. Church, S. Lee, E. A. L. Bagg et al., "A mouse model for the metabolic effects of the human fat mass and obesity associated FTO gene," PLoS Genetics, vol. 5, no. 8, Article ID e1000599, 2009.

47. C. Dina, D. Meyre, S. Gallina et al., "Variation in FTO contributes to childhood obesity and severe adult obesity," Nature Genetics, vol. 39, no. 6, pp. 724–726, 2007.

48. T. M. Frayling, N. J. Timpson, M. N. Weedon et al., "A common variant in the FTO gene is associated with body mass index and predisposes to childhood and adult obesity," Science, vol. 316, no. 5826, pp. 889–894, 2007.

49. S. Ereqat, A. Nasereddin, S. Cauchi, K. Azmi, Z. Abdeen, and R. Amin, "Association of a common variant in TCF7L2 gene with type 2 diabetes mellitus in the Palestinian population," Acta Diabetologica, vol. 47, no. 1, pp. S195–S198, 2010.

50. M. M. Zaharna, A. A. Abed, and F. A. Sharif, "Calpain-10 gene polymorphism in type 2 diabetes mellitus patients in the gaza strip," Medical Principles and Practice, vol. 19, no. 6, pp. 457–462, 2010.

51. I. Ezzidi, N. Mtiraoui, R. Nemr et al., "Variants within the calpain-10 gene and relationships with type 2 diabetes (T2DM) and T2DM-related traits among Tunisian Arabs," Diabetes and Metabolism, vol. 36, no. 5, pp. 357–362, 2010.

52. S. Mehri, B. Baudin, S. Mahjoub et al., "Angiotensin-converting enzyme insertion/ deletion gene polymorphism in a Tunisian healthy and acute myocardial infarction population," Genetic Testing and Molecular Biomarkers, vol. 14, no. 1, pp. 85–91, 2010.

53. T. P. They-They, K. Hamzi, M. T. Moutawafik, H. Bellayou, M. El Messal, and S. Nadifi, "Prevalence of angiotensin-converting enzyme, methylenetetrahydrofolate reductase, Factor v Leiden, prothrombin and apolipoprotein e gene polymorphisms in Morocco," Annals of Human Biology, vol. 37, no. 6, pp. 767–777, 2010.

54. M. Akra-Ismail, R. F. Makki, H. N. Chmaisse, A. Kazma, and N. K. Zgheib, "Association between angiotensin-converting enzyme insertion/deletion genetic polymorphism and hypertension in a sample of Lebanese patients," Genetic Testing and Molecular Biomarkers, vol. 14, no. 6, pp. 787–792, 2010.

55. R. Nemr, R. A. Salman, L. H. Jawad, E. A. Juma, S. H. Keleshian, and W. Y. Almawi, "Differential contribution of MTHFR C677T variant to the risk of diabetic nephropathy in Lebanese and Bahraini Arabs," Clinical Chemistry and Laboratory Medicine, vol. 48, no. 8, pp. 1091–1094, 2010.

56. K. K. Abu-Amero, C. A. Wyngaard, O. M. Al-Boudari, M. Kambouris, and N. Dzimiri, "Lack of association of lipoprotein lipase gene polymorphisms with coronary artery disease in the Saudi Arab population," Archives of Pathology and Laboratory Medicine, vol. 127, no. 5, pp. 597–600, 2003.

57. K. K. Abu-Amero, O. M. Al-Boudari, G. H. Mohamed, and N. Dzimiri, "The Glu27 genotypes of the Beta2-adrenergic receptor are predictors for severe coronary artery disease," BMC Medical Genetics, vol. 7, article 31, 2006.

58. K. Shiwaku, A. Nogi, E. Anuurad et al., "Difficulty in losing weight by behavioral intervention for women with Trp64Arg polymorphism of the β3-adrenergic receptor gene," International Journal of Obesity, vol. 27, no. 9, pp. 1028–1036, 2003.

59. A. Memisoglu, F. B. Hu, S. E. Hankinson et al., "Interaction between a peroxisome proliferator-activated receptor γ gene polymorphism and dietary fat intake in relation to body mass," Human Molecular Genetics, vol. 12, no. 22, pp. 2923–2929, 2003.

60. J. Luan, P. O. Browne, A. H. Harding et al., "Evidence for gene-nutrient interaction at the PPARγ locus," Diabetes, vol. 50, no. 3, pp. 686–689, 2001.

61. V. Lindi, U. Schwab, A. Louheranta et al., "Impact of the Pro12Ala polymorphism of the PPAR-γ2 gene on serum triacylglycerol response to n-3 fatty acid supplementation," Molecular Genetics and Metabolism, vol. 79, no. 1, pp. 52–60, 2003.

62. J. A. Moreno, J. López-Miranda, C. Marín et al., "The influence of the apolipoprotein E gene promoter (-219G/T) polymorphism on postprandial lipoprotein metabolism in young normolipemic males," Journal of Lipid Research, vol. 44, no. 11, pp. 2059–2064, 2003.

63. G. J. Petot, F. Traore, S. M. Debanne, A. J. Lerner, K. A. Smyth, and R. P. Friedland, "Interactions of apolipoprotein E genotype and dietary fat intake of healthy older persons during mid-adult life," Metabolism, vol. 52, no. 3, pp. 279–281, 2003.

64. B. J. Nicklas, R. E. Ferrell, L. B. Bunyard, D. M. Berman, K. E. Dennis, and A. P. Goldberg, "Effects of apolipoprotein E genotype on dietary-induced changes in high-density lipoprotein cholesterol in obese postmenopausal women," Metabolism, vol. 51, no. 7, pp. 853–858, 2002.

65. J. T. Bazzaz, M. Nazari, H. Nazem et al., "Apolipoprotein e gene polymorphism and total serum cholesterol level in Iranian population," Journal of Postgraduate Medicine, vol. 56, no. 3, pp. 173–175, 2010.

66. A. O. Akanji, C. G. Suresh, H. R. Fatania, R. Al-Radwan, and M. Zubaid, "Associations of apolipoprotein E polymorphism with low-density lipoprotein size and subfraction profiles in Arab patients with coronary heart disease," Metabolism, vol. 56, no. 4, pp. 484–490, 2007.

67. J. M. Ordovas, D. Corella, S. Demissie et al., "Dietary fat intake determines the effect of a common polymorphism in the hepatic lipase gene promoter on high-density lipoprotein metabolism: evidence of a strong dose effect in this gene-nutrient interaction in the Framingham study," Circulation, vol. 106, no. 18, pp. 2315–2321, 2002.

68. E. S. Tai, D. Corella, M. Deurenberg-Yap et al., "Dietary fat interacts with the -514C>T polymorphism in the hepatic lipase gene promoter on plasma lipid profiles in a Multiethnic Asian population: the 1998 Singapore National Health Survey," Journal of Nutrition, vol. 133, no. 11, pp. 3399–3408, 2003.

69. L. S. Rozek, T. S. Hatsukami, R. J. Richter et al., "The correlation of paraoxonase (PON1) activity with lipid and lipoprotein levels differs with vascular disease status," Journal of Lipid Research, vol. 46, no. 9, pp. 1888–1895, 2005.

70. G. P. Jarvik, L. S. Rozek, V. H. Brophy et al., "Paraoxonase (PON1) phenotype is a better predictor of vascular disease than is PON1192 or PON155 genotype," Arteriosclerosis, Thrombosis, and Vascular Biology, vol. 20, no. 11, pp. 2441–2447, 2000.

71. B. Agachan, H. Yilmaz, Z. Karaali, and T. Isbir, "Paraoxonase 55 and 192 polymorphism and its relationship to serum paraoxonase activity and serum lipids in Turkish patients with non-insulin dependent diabetes mellitus," Cell Biochemistry and Function, vol. 22, no. 3, pp. 163–168, 2004.

72. L. G. Costa, A. Vitalone, T. B. Cole, and C. E. Furlong, "Modulation of paraoxonase (PON1) activity," Biochemical Pharmacology, vol. 69, no. 4, pp. 541–550, 2005.

73. I. Nabipour, M. Amiri, S. R. Imami et al., "The metabolic syndrome and nonfatal ischemic heart disease; a population-based study," International Journal of Cardiology, vol. 118, no. 1, pp. 48–53, 2007.

74. M. Nsour, Z. Mahfoud, M. N. Kanaan, and A. Balbeissi, "Prevalence and predictors of non-fatal myocardial infarction in Jordan," Eastern Mediterranean Health Journal, vol. 14, no. 4, pp. 818–830, 2008.

75. M. M. Al-Nozha, M. R. Arafah, Y. Y. Al-Mazrou et al., "Coronary artery disease in Saudi Arabia," Saudi Medical Journal, vol. 25, no. 9, pp. 1165–1171, 2004.

76. H. Ben Romdhane, R. Khaldi, A. Oueslati, and H. Skhiri, "Transition épidémiologique et transition alimentaire et nutritionnelle en Tunisie," Options Méditerranéennes B, vol. 41, 2002.

77. A. Al-Jishi and P. Mohan, "Profile of stroke in Bahrain," Neurosciences, vol. 5, no. 1, pp. 30–34, 2000.

78. A. A. Ahangar, S. B. A. Vaghefi, and M. Ramaezani, "Epidemiological evaluation of stroke in Babol, Northern Iran (2001–2003)," European Neurology, vol. 54, no. 2, pp. 93–97, 2005.

79. N. U. A. M. A. Abdul-Ghaffar, M. R. El-Sonbaty, M. S. El-Din Abdul-Baky, A. A. Marafie, and A. M. Al-Said, "Stroke in Kuwait: a three-year prospective study," Neuroepidemiology, vol. 16, no. 1, pp. 40–47, 1997.

80. K. Radhakrishnan, P. P. Ashok, R. Sridharan, and M. A. El-Mangoush, "Incidence and pattern of cerebrovascular diseases in Benghazi, Libya," Journal of Neurology Neurosurgery and Psychiatry, vol. 49, no. 5, pp. 519–523, 1986.

81. W. M. Sweileh, A. F. Sawalha, S. M. Al-Aqad, et al., "The epidemiology of stroke in northern palestine: a 1-year, hospital-based study," Journal of Stroke and Cerebro-vascular Diseases, vol. 17, no. 6, pp. 406–411, 2008.

82. A. Hamad, A. Hamad, T. E. O. Sokrab, S. Momeni, B. Mesraoua, and A. Lingren, "Stroke in Qatar: a one-year, hospital-based study," Journal of Stroke and Cerebro-vascular Diseases, vol. 10, no. 5, pp. 236–241, 2001.

83. S. Al-Rajeh, E. B. Larbi, O. Bademosi et al., "Stroke register: experience from the Eastern Province of Saudi Arabia," Cerebrovascular Diseases, vol. 8, no. 2, pp. 86–89, 1998.

84. A. Yahia-Berrouiguet, M. Benyoucef, K. Meguenni, and M. Brouri, "Prevalence of cardiovascular risk factors: a survey at Tlemcen (Algeria)," Medecine des Maladies Metaboliques, vol. 3, no. 3, pp. 313–319, 2009.

85. F. I. Al-Zurba, "Latest studies clarify state of health in Bahrain," Diabetes Voice, vol. 46, pp. 28–31, 2001.

86. M. M. Ibrahim, H. Rizk, L. J. Appel et al., "Hypertension prevalence, awareness, treatment, and control in Egypt: results from the Egyptian National Hypertension Project (NHP)," Hypertension, vol. 26, no. 6, pp. 886–890, 1995.

87. O. M. Galal, "The nutrition transition in Egypt: obesity, undernutrition and the food consumption context," Public Health Nutrition, vol. 5, no. 1, pp. 141–148, 2002.

88. N. Sarraf-Zadegan, M. Boshtam, S. Mostafavi, and M. Rafiei, "Prevalence of hy-pertension and associated risk factors in Isfahan, Islamic Republic of Iran," Eastern Mediterranean Health Journal, vol. 5, no. 5, pp. 992–1001, 1999.

89. A. A. Haghdoost, B. Sadeghirad, and M. Rezazadehkermani, "Epidemiology and heterogeneity of hypertension in Iran: a systematic review," Archives of Iranian Medicine, vol. 11, no. 4, pp. 444–452, 2008.

90. WHO, "STEPwise surveillance. Non-communicable diseases risk factors. STEP-wise data from selected countries in the Eastern Mediterranean Region, 2003–2007," 2007, http://www.emro.who.int/ncd/risk_factors.htm#physical.

91. M. Zindah, A. Belbeisi, H. Walke, and A. H. Mokdad, "Obesity and diabetes in Jor-dan: findings from the behavioral risk factor surveillance system, 2004," Preventing Chronic Disease, vol. 5, pp. 1–8, 2008.

92. A.-M. Sibai, O. Obeid, M. Batal, N. Adra, D. E. Khoury, and N. Hwalla, "Prevalence and correlates of metabolic syndrome in an adult Lebanese population," CVD Pre-vention and Control, vol. 3, no. 2, pp. 83–90, 2008.

93. M. A. Tazi, S. Abir-Khalil, N. Chaouki et al., "Prevalence of the main cardiovascular risk factors in Morocco: results of a National Survey, 2000," Journal of Hyperten-sion, vol. 21, no. 5, pp. 897–903, 2003.

94. A. A. Hasab, A. Jaffer, and Z. Hallaj, "Blood pressure patterns among the Omani population," Eastern Mediterranean Health Journal, vol. 5, no. 1, pp. 46–54, 1999.

95. H. F. Abdul-Rahim, G. Holmboe-Ottesen, L. C. M. Stene et al., "Obesity in a rural and an urban Palestinian West Bank population," International Journal of Obesity, vol. 27, no. 1, pp. 140–146, 2003.

96. A. Bener, J. Al-Suwaidi, K. Al-Jaber, S. Al-Marri, M. H. Dagash, and I. E. Elbagi, "The prevalence of hypertension and its associated risk factors in a newly developed country," Saudi Medical Journal, vol. 25, no. 7, pp. 918–922, 2004.

97. M. M. Al-Nozha, M. Abdullah, M. R. Arafah et al., "Hypertension in Saudi Arabia," Saudi Medical Journal, vol. 28, no. 1, pp. 77–84, 2007.

98. W. Maziak, S. Rastam, F. Mzayek, K. D. Ward, T. Eissenberg, and U. Keil, "Cardiovascular health among adults in Syria: a model from developing countries," Annals of Epidemiology, vol. 17, no. 9, pp. 713–720, 2007.

99. H. M. Sonmez, O. Basak, C. Camci et al., "The epidemiology of elevated blood pressure as an estimate for hypertension in Aydin, Turkey," Journal of Human Hypertension, vol. 13, no. 6, pp. 399–404, 1999.

100. L. M. Baynouna, A. D. Revel, N. J. Nagelkerke et al., "High prevalence of the cardiovascular risk factors in Al-Ain, United Arab Emirates. An emerging health care priority," Saudi Medical Journal, vol. 29, no. 8, pp. 1173–1178, 2008.

101. A. A. Gunaid and A. M. Assabri, "Prevalence of type 2 diabetes and other cardiovascular risk factors in a semirural area in Yemen," La Revue de Santé de la Méditerranée Orientale, vol. 14, no. 1, pp. 42–56, 2008.

102. B. Motlagh, M. O'Donnell, and S. Yusuf, "Prevalence of cardiovascular risk factors in the middle east: a systematic review," European Journal of Cardiovascular Prevention and Rehabilitation, vol. 16, no. 3, pp. 268–280, 2009.

103. A. Mehio Sibai, L. Nasreddine, A. H. Mokdad, N. Adra, M. Tabet, and N. Hwalla, "Nutrition transition and cardiovascular disease risk factors in Middle East and North Africa countries: reviewing the evidence," Annals of Nutrition and Metabolism, vol. 57, no. 3-4, pp. 193–203, 2010.

104. Y. Khader, A. Bateiha, M. El-Khateeb, A. Al-Shaikh, and K. Ajlouni, "High prevalence of the metabolic syndrome among Northern Jordanians," Journal of Diabetes and Its Complications, vol. 21, no. 4, pp. 214–219, 2007.

105. I. Al Rashdan and Y. Al Nesef, "Prevalence of overweight, obesity, and metabolic syndrome among adult Kuwaitis: results from community-based national survey," Angiology, vol. 61, no. 1, pp. 42–48, 2010.

106. M. Rguibi and R. Belahsen, "Metabolic syndrome among Moroccan Sahraoui adult women," American Journal of Human Biology, vol. 16, no. 5, pp. 598–601, 2004.

107. J. A. Al-Lawati and P. Jousilahti, "Prevalence of metabolic syndrome in Oman using the International Diabetes Federation's criteria," Saudi Medical Journal, vol. 27, no. 12, pp. 1925–1926, 2006.

108. M. Musallam, A. Bener, M. Zirie et al., "Metabolic syndrome and its components among Qatari population," International Journal of Food Safety, Nutrition and Public Health, vol. 1, no. 1, pp. 88–102, 2008.

109. A. Bener, M. Zirie, M. Musallam, Y. S. Khader, and A. O. Al-Hamaq, "Prevalence of metabolic syndrome according to adult treatment panel III and international diabetes federation criteria: a population-based study," Metabolic Syndrome and Related Disorders, vol. 7, no. 3, pp. 221–230, 2009.

110. M. M. Al-Nozha, A. Al-Khadra, M. R. Arafah et al., "Metabolic syndrome in Saudi Arabia," Saudi Medical Journal, vol. 26, no. 12, pp. 1918–1925, 2005.

111. R. Bouguerra, L. Ben Salem, H. Alberti et al., "Prevalence of metabolic abnormalities in the Tunisian adults: a population based study," Diabetes and Metabolism, vol. 32, no. 3, pp. 215–221, 2006.

112. M. Malik and S. A. Razig, "The prevalence of the metabolic syndrome among the multiethnic population of the United Arab Emirates: a report of a national survey," Metabolic Syndrome and Related Disorders, vol. 6, no. 3, pp. 177–186, 2008.

113. P. M. Kearney, M. Whelton, K. Reynolds, P. K. Whelton, and J. He, "Worldwide prevalence of hypertension: a systematic review," Journal of Hypertension, vol. 22, no. 1, pp. 11–19, 2004.

114. T. A. Gaziano, A. Bitton, S. Anand, S. Abrahams-Gessel, and A. Murphy, "Growing epidemic of coronary heart disease in low- and middle-income countries," Current Problems in Cardiology, vol. 35, no. 2, pp. 72–115, 2010.

115. R. M. Mabry, M. M. Reeves, E. G. Eakin, and N. Owen, "Gender differences in prevalence of the metabolic syndrome in Gulf Cooperation Council Countries: a systematic review," Diabetic Medicine, vol. 27, no. 5, pp. 593–597, 2010.

116. J. Tran, M. Mirzaei, L. Anderson, and S. R. Leeder, "The epidemiology of stroke in the Middle East and North Africa," Journal of the Neurological Sciences, vol. 295, no. 1-2, pp. 38–40, 2010.

117. G. P. Page, J. W. Edwards, S. Barnes, R. Weindruch, and D. B. Allison, "A design and statistical perspective on microarray gene expression studies in nutrition: the need for playful creativity and scientific hard-mindedness," Nutrition, vol. 19, no. 11-12, pp. 997–1000, 2003.

118. M. F. Fenech, "Dietary reference values of individual micronutrients and nutriomes for genome damage prevention: current status and a road map to the future," American Journal of Clinical Nutrition, vol. 91, no. 5, pp. 1438S–1454S, 2010.

119. S. Kimball, G. E. H. Fuleihan, and R. Vieth, "Vitamin D: a growing perspective," Critical Reviews in Clinical Laboratory Sciences, vol. 45, no. 4, pp. 339–414, 2008.

120. R. Chesney, "Metabolic bone disease," in Nelson Textbook of Pediatrics, R. E. Behrman, R. Kliegman, and H. B. Jenson, Eds., pp. 2132–2138, WB Saunders, Philadelphia, Pa, USA, 16th edition, 2000.

121. S. Wang, "Epidemiology of vitamin D in health and disease," Nutrition Research Reviews, vol. 22, no. 2, pp. 188–203, 2009.

122. C. F. Garland, F. C. Garland, E. D. Gorham et al., "The role of vitamin D in cancer prevention," American Journal of Public Health, vol. 96, no. 2, pp. 252–261, 2006.

123. M. Janner, P. Ballinari, P. E. Mullis, and C. E. Flück, "High prevalence of vitamin D deficiency in children and adolescents with type 1 diabetes," Swiss Medical Weekly, vol. 140, article w13091, 2010.

124. J. S. Adams and M. Hewison, "Update in vitamin D," Journal of Clinical Endocrinology and Metabolism, vol. 95, no. 2, pp. 471–478, 2010.

125. S. R. Kreiter, R. P. Schwartz, H. N. Kirkman, P. A. Charlton, A. S. Calikoglu, and M. L. Davenport, "Nutritional rickets in African American breast-fed infants," Journal of Pediatrics, vol. 137, no. 2, pp. 153–157, 2000.

126. N. G. Jablonski and G. Chaplin, "The evolution of human skin coloration," Journal of Human Evolution, vol. 39, no. 1, pp. 57–106, 2000.

127. P. Lee, J. R. Greenfield, M. J. Seibel, J. A. Eisman, and J. R. Center, "Adequacy of vitamin D replacement in severe deficiency is dependent on body mass index," American Journal of Medicine, vol. 122, no. 11, pp. 1056–1060, 2009.

128. M. H. Gannagé-Yared, G. Maalouf, S. Khalife et al., "Prevalence and predictors of vitamin D inadequacy amongst Lebanese osteoporotic women," British Journal of Nutrition, vol. 101, no. 4, pp. 487–491, 2009.

129. K. Holvik, H. E. Meyer, E. Haug, and L. Brunvand, "Prevalence and predictors of vitamin D deficiency in five immigrant groups living in Oslo, Norway: the Oslo immigrant health study," European Journal of Clinical Nutrition, vol. 59, no. 1, pp. 57–63, 2005.

130. G. I. Baroncelli, A. Bereket, M. El Kholy et al., "Rickets in the Middle East: role of environment and genetic predisposition," Journal of Clinical Endocrinology and Metabolism, vol. 93, no. 5, pp. 1743–1750, 2008.

131. S. K. Ames, K. J. Ellis, S. K. Gunn, K. C. Copeland, and S. A. Abrams, "Vitamin D receptor gene Fok1 polymorphism predicts calcium absorption and bone mineral density in children," Journal of Bone and Mineral Research, vol. 14, no. 5, pp. 740–746, 1999.

132. H. Arai, K. I. Miyamoto, Y. Taketani et al., "A vitamin D receptor gene polymorphism in the translation initiation codon: effect on protein activity and relation to bone mineral density in Japanese women," Journal of Bone and Mineral Research, vol. 12, no. 6, pp. 915–921, 1997.

133. C. D. Davis, "Vitamin D and cancer: current dilemmas and future research needs," American Journal of Clinical Nutrition, vol. 88, no. 2, 2008.

134. M. A. Abdullah, H. S. Salhi, L. A. Bakry et al., "Adolescent rickets in Saudi Arabia: a rich and sunny country," Journal of Pediatric Endocrinology and Metabolism, vol. 15, no. 7, pp. 1017–1025, 2002.

135. H. Arai, K. I. Miyamoto, M. Yoshida et al., "The polymorphism in the caudal-related homeodomain protein Cdx-2 binding element in the human vitamin D receptor gene," Journal of Bone and Mineral Research, vol. 16, no. 7, pp. 1256–1264, 2001.

136. A. V. Krishnan, S. Swami, J. Moreno, R. B. Bhattacharyya, D. M. Peehl, and D. Feldman, "Potentiation of the growth-inhibitory effects of vitamin D in prostate cancer by genistein," Nutrition Reviews, vol. 65, no. 8, pp. S121–123, 2007.

137. A. Ali, G. Fathy, H. Fathy, and N. Abd El-Ghaffar, "Epidemiology of iron deficiency anaemia: effect of physical growth in primary school children, the importance of hookworms," International Journal of Academic Research, vol. 3, pp. 495–500, 2011.

138. W. Burke, G. Imperatore, and M. Reyes, "Iron deficiency and iron overload: effects of diet and genes," Proceedings of the Nutrition Society, vol. 60, no. 1, pp. 73–80, 2001.

139. Chung and R. Wessling-Resnick, "Lessons learned from genetic and nutritional iron deficiencies," Nutrition Reviews, vol. 62, no. 5, pp. 212–215, 2004.

140. J. E. Levy, L. K. Montross, and N. C. Andrews, "Genes that modify the hemochromatosis phenotype in mice," Journal of Clinical Investigation, vol. 105, no. 9, pp. 1209–1216, 2000.

141. L. B. Bailey and J. F. Gregory, "Folate metabolism and requirements," Journal of Nutrition, vol. 129, no. 4, pp. 779–782, 1999.

142. S. W. Choi and J. B. Mason, "Folate and carcinogenesis: an integrated scheme," Journal of Nutrition, vol. 130, no. 2, pp. 129–132, 2000.

143. M. Moussavi, R. Heidarpour, A. Aminorroaya, Z. Pournaghshband, and M. Amini, "Prevalence of vitamin D deficiency in Isfahani high school students in 2004," Hormone Research, vol. 64, no. 3, pp. 144–148, 2005.

144. A. Batieha, Y. Khader, H. Jaddou et al., "Vitamin D status in Jordan: dress style and gender discrepancies," Annals of Nutrition and Metabolism, vol. 58, no. 1, pp. 10–18, 2011.

145. G. El-Hajj Fuleihan, M. Nabulsi, M. Choucair et al., "Hypovitaminosis D in healthy schoolchildren," Pediatrics, vol. 107, article E53, 2001.

146. G. El-Hajj Fuleihan, M. Nabulsi, H. Tamim et al., "Effect of vitamin D replacement on musculoskeletal parameters in school children: a randomized controlled trial," The Journal of Clinical Endocrinology and Metabolism, vol. 91, no. 2, pp. 405–412, 2006.

147. A. Arabi, R. El Rassi, and G. El-Hajj Fuleihan, "Hypovitaminosis D in developing countries-prevalence, risk factors and outcomes," Nature Reviews Endocrinology, vol. 6, no. 10, pp. 550–561, 2010.

148. A. M. Siddiqui and H. Z. Kamfar, "Prevalence of vitamin D deficiency rickets in adolescent school girls in Western region, Saudi Arabia," Saudi Medical Journal, vol. 28, no. 3, pp. 441–444, 2007.

149. A. O. Musaiger, "Iron deficiency anaemia among children and pregnant women in the arab gulf countries: the need for action," Nutrition and Health, vol. 16, no. 3, pp. 161–171, 2002.

150. K. Bagchi, "Iron deficiency anaemia—an old enemy," La Revue de Santé de la Méditerranée Orientale, vol. 10, no. 6, pp. 754–760, 2004.

151. W. Y. Almawi, R. R. Finan, H. Tamim, J. L. Daccache, and N. Irani-Hakime, "Differences in the Frequency of the C677T Mutation in the Methylenetetrahydrofolate Reductase (MTHFR) Gene Among the Lebanese Population," American Journal of Hematology, vol. 76, no. 1, pp. 85–87, 2004.

152. A. M. T. Engbersen, D. G. Franken, G. H. J. Boers, E. M. B. Stevens, F. J. M. Trijbels, and H. J. Blom, "Thermolabile 5,10-methylenetetrahydrofolate reductase as a cause of mild hyperhomocysteinemia," American Journal of Human Genetics, vol. 56, no. 1, pp. 142–150, 1995.

153. N. M. van der Put and H. J. Blom, "Neural tube defects and a disturbed folate dependent homocysteine metabolism," European Journal of Obstetrics Gynecology and Reproductive Biology, vol. 92, no. 1, pp. 57–61, 2000.

154. H. J. Huh, H. S. Chi, E. H. Shim, S. Jang, and C. J. Park, "Gene-nutrition interactions in coronary artery disease: correlation between the MTHFR C677T polymorphism and folate and homocysteine status in a Korean population," Thrombosis Research, vol. 117, no. 5, pp. 501–506, 2006.

155. J. C. Chambers, H. Ireland, E. Thompson et al., "Methylenetetrahydrofolate reductase 677 C→T mutation and coronary heart disease risk in UK Indian Asians," Arteriosclerosis, Thrombosis, and Vascular Biology, vol. 20, no. 11, pp. 2448–2452, 2000.

156. S. A. Ross and L. Poirier, "Proceedings of the Trans-HHS Workshop: diet, DNA methylation processes and health," Journal of Nutrition, vol. 132, no. 8, 2002.

157. J. Ma, M. J. Stampfer, E. Giovannucci et al., "Methylenetetrahydrofolate reductase polymorsphism, dietary interactions, and risk of colorectal cancer," Cancer Research, vol. 57, no. 6, pp. 1098–1102, 1997.

158. Z. Abdollahi, I. Elmadfa, A. Djazayeri et al., "Folate, vitamin B12 and homocysteine status in women of childbearing age: baseline data of folic acid wheat flour fortification in Iran," Annals of Nutrition and Metabolism, vol. 53, no. 2, pp. 143–150, 2008.

159. S. L. Tokgözoğlu, M. Alikaşifoğlu, I. Ünsal et al., "Methylene tetrahydrofolate reductase genotype and the risk and extent of coronary artery disease in a population with low plasma folate," Heart, vol. 81, no. 5, pp. 518–522, 1999.

160. B. Christensen, P. Frosst, S. Lussier-Cacan et al., "Correlation of a common mutation in the methylenetetrahydrofolate reductase gene with plasma homocysteine in patients with premature coronary artery disease," Arteriosclerosis, Thrombosis, and Vascular Biology, vol. 17, no. 3, pp. 569–573, 1997.

161. A. H. Messika, D. N. Kaluski, E. Lev et al., "Nutrigenetic impact of daily folate intake on plasma homocysteine and folate levels in patients with different methylenetetrahydrofolate reductase genotypes," European Journal of Cardiovascular Prevention and Rehabilitation, vol. 17, no. 6, pp. 701–705, 2010.

162. A. Wilson, R. Platt, Q. Wu et al., "A common variant in methionine synthase reductase combined with low cobalamin (Vitamin B12) increases risk for spina bifida," Molecular Genetics and Metabolism, vol. 67, no. 4, pp. 317–323, 1999.

163. L. E. Cahill, B. Fontaine-Bisson, and A. El-Sohemy, "Functional genetic variants of glutathione S-transferase protect against serum ascorbic acid deficiency," American Journal of Clinical Nutrition, vol. 90, no. 5, pp. 1411–1417, 2009.

164. L. E. Cahill and A. El-Sohemy, "Haptoglobin genotype modifies the association between dietary vitamin C and serum ascorbic acid deficiency," American Journal of Clinical Nutrition, vol. 92, no. 6, pp. 1494–1500, 2010.

165. J. M. Zingg, A. Azzi, and M. Meydani, "Genetic polymorphisms as determinants for disease-preventive effects of vitamin E," Nutrition Reviews, vol. 66, no. 7, pp. 406–414, 2008.

166. M. C. Zillikens, J. B. J. Van Meurs, F. Rivadeneira et al., "Interactions between dietary vitamin E intake and SIRT1 genetic variation influence body mass index," American Journal of Clinical Nutrition, vol. 91, no. 5, pp. 1387–1393, 2010.

167. H. Hamamy and A. Alwan, "Genetic disorders and congenital abnormalities: strategies for reducing the burden in the Region," Eastern Mediterranean Health Journal, vol. 3, no. 1, pp. 123–132, 1997.

PART II

CLINICAL NUTRITIONAL
BIOCHEMISTRY

CHAPTER 3

Consumption of Fructose and High Fructose Corn Syrup Increase Postprandial Triglycerides, LDL-Cholesterol, and Apolipoprotein-B in Young Men and Women

KIMBER L. STANHOPE, ANDREW A. BREMER,
VALENTINA MEDICI, KATSUYUKI NAKAJIMA, YASUKI ITO,
TAKAMITSU NAKANO, GUOXIA CHEN, TAK HOU FONG,
VIVIEN LEE, ROSEANNE I. MENORCA, NANCY L. KEIM,
AND PETER J. HAVEL

In epidemiological studies, consumption of sugar and/or sugar-sweetened beverages has been linked to the presence of unfavorable lipid levels (1–5), insulin resistance (6, 7), fatty liver (8, 9), type 2 diabetes (10–12), cardiovascular disease (13), and metabolic syndrome (14). We have recently reported that consumption of fructose-sweetened beverages at 25% of energy requirements (E) increased visceral adipose deposition and de novo lipogenesis, produced dyslipidemia, and decreased glucose tolerance/insulin sensitivity in older, overweight/obese men and women, whereas consumption of glucose-sweetened beverages did not (15). Because the com-

Consumption of Fructose and High Fructose Corn Syrup Increase Postprandial Triglycerides, LDL-Cholesterol, and Apolipoprotein-B in Young Men and Women. © Stanhope KL, Bremer AA, Medici V, Nakajima K, Ito Y, Nakano T, Chen G, Fong TH, Lee V, Menorca RI, Keim NL, and Havel PJ. The Journal of Clinical Endocrinology & Metabolism 96,10 (2011). DOI: http://dx.doi.org/10.1210/jc.2011-1251. Reprinted with permission from the authors.

monly consumed sugars, sucrose and high-fructose corn syrup (HFCS), are composed of 50–55% fructose, these results provide a potential mechanistic explanation for the associations between sugar consumption and metabolic disease. However, the adverse metabolic effects of fructose consumption observed in the older, overweight/obese population (15) may not occur in a younger, leaner population.

Authors of three recent reviews have concluded that long-term sugar intakes as high as 25–50% E have no adverse effects with respect to components of metabolic syndrome (16) and that fructose consumption up to 140 g/d does not result in biologically relevant increases of fasting or post-prandial triglycerides (TG) in healthy, normal-weight (17), or overweight or obese (18) humans. These reviews (16, 17) are cited in the Report of the Dietary Guidelines Advisory Committee on the Dietary Guidelines for Americans 2010, released June of 2010, in which a maximal intake level of 25% or less of total energy from added sugars is suggested (19). However, in August of 2009, the American Heart Association Nutrition Committee recommended that women consume no more than 100 kcal/d and men consume no more than 150 kcal/d of added sugar (20). This equates to differences between the two guidelines of 400 kcal/d for women consuming 2000 kcal/d and 525 kcal/d for men consuming 2500 kcal/d. To address this discrepancy, we compared the effects of consuming 25%E as glucose, fructose or HFCS for 2 weeks on risk factors for cardiovascular disease in young adults.

3.1 MATERIALS AND METHODS

The subjects who participated in this study are a subgroup of participants from an ongoing 5-yr National Institutes of Health-funded investigation in which a total of eight experimental groups (n = 25/group) will be studied. The objectives include comparing the metabolic effects of fructose, glucose, and HFCS consumption at 25% E and to compare the metabolic effects of fructose and HFCS consumption at 0, 10, 17.5, and 25% E. The results reported in this paper are from the first 48 subjects to complete the study protocol in the experimental groups consuming 25% E as glucose, fructose, or HFCS (n = 16/group). Participants were recruited through an internet listing (Craigslist.com) and underwent telephone and in-person interviews with medical history, complete blood count, and serum bio-chemistry panel to assess eligibility. Inclusion criteria included age 18–40 yr and body mass index (BMI) 18–35 kg/m^2 with a self-report of stable body weight during the prior 6 months. Exclusion criteria included diabe-tes (fasting glucose >125 mg/dl), evidence of renal or hepatic disease, fast-ing plasma TG greater than 400 mg/dl, hypertension (>140/90 mm Hg), or surgery for weight loss. Individuals who smoked, habitually ingested more than two alcoholic beverages per day, exercised more than 3.5 h/wk at a level more vigorous than walking, or used thyroid, lipid-lowering, glucose-lowering, antihypertensive, antidepressant, or weight loss medi-cations were also excluded. The University of California, Davis, Institu-tional Review Board approved the experimental protocol for this study, and subjects provided written informed consent to participate.

For the 5 wk before study, subjects were asked to limit daily consump-tion of sugar-containing beverages to one 8-oz serving of fruit juice. Fifty-five subjects were enrolled in the experimental groups consuming 25% E as glucose, fructose, or HFCS. Four subjects withdrew due to unwilling-ness to comply with the study protocol (two in the HFCS group, two before group assignment), and two were withdrawn due to medical conditions not apparent during screening (HFCS and glucose group). The samples from one subject (HFCS group) who completed the study protocol were not analyzed because of illness during the 24-h serial blood collection. The experimental groups were matched for gender (nine men, seven women/group), BMI, fasting TG, cholesterol, high-density lipoprotein (HDL) and insulin concentrations. The subjects and University of California, Davis, Clinical Research Center (CCRC) and technical personnel were blinded to the sugar assignments.

This was a parallel-arm, diet intervention study with three phases: 1) a 3.5-d inpatient baseline period during which subjects resided at the CCRC; 2) a 12-d outpatient intervention period; and 3) a 3.5-d inpatient interven-tion period at the CCRC. During d 2 and 3 of the baseline and intervention inpatient periods, subjects consumed energy-balanced meals consisting of conventional foods. Daily energy requirements were calculated by the Mifflin equation (21) with adjustment of 1.3 for activity on the days of the 24-h serial blood collections, and adjustment of 1.5 for the other days. The baseline diet contained 55% E mainly as complex carbohydrate, 30% fat, and 15% protein. The intervention inpatient meals were as identical as possible to baseline meals, excepting the carbohydrate component consisted of 25% E as glucose-, fructose-, or HFCS-sweetened beverages and 30% E as complex carbohydrate. Sugar-sweetened beverages were provided to subjects as three daily servings consumed with meals and were flavored with an unsweetened drink mix (Kool-Aid; Kraft Foods, Northfield, IL). The timing of inpatient meal service and the energy distribution were: breakfast, 0900 h (25%); lunch, 1300 h (35%); dinner, 1800 h (40%).

During the 12-d outpatient phase of the study, the subjects were provided with and instructed to drink three servings of sugar-sweetened beverage per day (one per meal), to consume their usual diet, and to not consume other sugar-containing beverages, including fruit juice. To monitor compliance, the sugar-sweetened beverages contained a biomarker (riboflavin), which was measured fluorometrically in urine samples collected at the time of beverage pickup. These measurements indicated that the three groups of subjects were comparably compliant.

Twenty-four-hour serial blood collections were conducted on the third day of the baseline (0 wk) and intervention (2 wk) inpatient periods. Three fasting blood samples were collected at 0800, 0830, and 0900 h. Twenty-nine postprandial blood samples were collected at 30- to 60-min intervals from 0930 until 0800 h the next morning. Additional 6-ml samples were collected at the fasting time points, 0800, 0830, and 0900 h and also at 2200, 2300, and 2400 h, the period during which TG concentrations peaked during our previous study (15). The additional plasma from the three fasting samples was pooled, as was that from the three late-evening postprandial samples; multiple aliquots of each pooled sample were stored at −80 C.

3.1.1 ANALYSES

Primary outcomes include fasting TG, 24-h TG incremental area under the curve (AUC), late-evening postprandial TG concentrations, and fasting LDL, non-HDL-cholesterol (-C), apolipoprotein (apo)B concentrations, and the apoB to apoAI ratio. Secondary outcomes included body weight, fasting HDL, postprandial LDL, non-HDL-C, apoB, remnant lipoprotein-cholesterol (RLP)-C and RLP-TG, and fasting and postprandial small dense LDL-cholesterol (sdLDL-C). Fasting concentrations, 24-h AUC, and postmeal peaks for glucose and insulin, and homeostasis model assessment insulin resistance index (HOMA-IR) are presented in the online supplement, published on The Endocrine Society's Journals Online web site at http://jcem.endojournals.org. Fasting measures were conducted on samples collected or pooled from the 0800, 0830, and 0900 h time points, and postprandial measures were conducted on samples collected or pooled from the 2200, 2300, and 2400 h time points. Lipid and lipoprotein concentrations (total cholesterol, HDL, TG, apoB, apoA1) were determined with a Polychem chemistry analyzer (PolyMedCo, Inc., Cortlandt Manor, NY). LDL concentrations were determined by direct homogenous assay using detergents (Denka Seiken, Tokyo, Japan) (22) and sdLDL-C concentrations were quantified using the sdLDL-C-EX"SEIKEN" homogeneous assay kit (Denka Seiken) (23). RLP concentrations were quantified with an immunoseparation assay (24). Glucose was measured with an automated glucose analyzer (YSI, Inc., Yellow Springs, OH), and insulin by RIA (Millipore, St. Charles, MO).

The incremental 24-h area AUC was calculated for TG, glucose, and insulin by the trapezoidal method. Glucose and insulin postmeal peaks were assessed as the mean amplitudes of the three postmeal peaks; specifically the peak postmeal value minus the premeal value was averaged for breakfast, lunch, and dinner for each subject. The absolute change (Δ from 2 wk when 25% E sugar/30% E complex carbohydrate was consumed compared with 0 wk when 55% E complex carbohydrate was consumed) for each outcome was analyzed with SAS 9.2 (SAS, Cary, NC) in a mixed procedures (PROC MIXED) model with sugar and gender as factors, and BMI, the change (2 to 0 wk) in body weight (ΔBW), and outcome concentration at baseline (outcome$_B$) as continuous covariables. ΔBW and outcome$_B$ were removed if they did not improve the precision of the model. Significant differences ($P < 0.05$) among the three sugars were identified by the Tukey's multiple comparisons test. Outcomes that were significantly affected by 2 wk of glucose, fructose, or HFCS consumption were identified as least squares means (LS means) of the change significantly different than zero. Primary outcomes were also analyzed with BMI as a factor (BMI <25 m/kg^2 vs. >25 m/kg^2). Data are presented as mean ± sem.

3.2 RESULTS

There were no significant differences among the three experimental groups in anthropomorphic (Table 1) or outcome measures at baseline (Tables 2 and 3 and Supplemental Table 1). Body weight (Table 3) and blood pressure (data not shown) were not affected by 2 wk consumption of glucose, fructose, or HFCS.

TABLE 1. Subjects' baseline anthropomorphic and metabolic parameters

Parameter	Glucose (n = 16)	Fructose (n = 16)	HFCS (n = 16)
Age (yr)	27.0 ± 7.2	28.0 ± 6.8	27.8 ± 7.6
Weight (kg)	76.8 ± 14.1	76.8 ± 10.6	74.3 ± 14.9
BMI (kg/m2)	26.2 ± 3.6	25.4 ± 3.8	24.9 ± 4.8
Waist circumference (cm)	80.6 ± 10.4	77.8 ± 9.6	78.0 ± 10.8
Body fat (%)	28.0 ± 9.3	26.7 ± 11.8	25.0 ± 10.1
TG (mmol/liter)	1.2 ± 0.5	1.2 ± 0.4	1.3 ± 0.6
Total cholesterol (mmol/liter)	4.5 ± 0.8	3.9 ± 0.8	4.1 ± 0.8
HDL-C (mmol/liter)	1.2 ± 0.4	1.2 ± 0.4	1.2 ± 0.4
Insulin (pmol/liter)	97.9 ± 30.4	102.8 ± 86.4	89.1 ± 31.6

P > 0.05 for differences among groups at baseline for all parameters, PROC MIXED ANOVA. Mean ± sd.

3.2.1 PRIMARY OUTCOMES: COMPARING GLUCOSE, FRUCTOSE, AND HFCS WITH COMPLEX CARBOHYDRATE CONSUMPTION

Table 2 presents the primary outcomes during consumption of complex carbohydrate at baseline (0 wk) and at the end of the 2-wk sugar interventions. The 24-h TG profiles during baseline and the end of the 2-wk inter-vention are shown in Fig. 1, A–C. The 24-h TG AUC (Fig. 2A) was signifi-cantly increased compared with baseline (LS means of Δ different from zero) in subjects consuming fructose (+4.7 ± 1.2 mmol/liter × 24 h, P = 0.0032) and HFCS (+1.8 ± 1.4 mmol/liter × 24 h, P = 0.035), whereas it tended to decrease during consumption of glucose (−1.9 ± 0.9 mmol/liter × 24 h, P = 0.14). The consumption of all three sugars resulted in a late-evening TG peak between 2200 and 2400 h that was not apparent when complex carbo-hydrate was consumed (Fig. 1, A–C). The late-evening peaks (Fig. 2B) were significantly increased compared with baseline during consumption of fruc-tose (+0.59 ± 0.11 mmol/liter, P < 0.0001) and HFCS (+0.46 ± 0.082 mmol/liter, P < 0.0001) but not by glucose (+0.22 ± 0.10 mmol/liter, P = 0.077). All three sugars tended to increase fasting TG, but this was significant only in the group consuming glucose (Fig. 2C). Fasting LDL-C concentrations (Fig. 3A) were increased during consumption of fructose (+0.29 ± 0.082 mmol/liter, P = 0.0023) and HFCS (+0.42 ± 0.11 mmol/liter, P < 0.0001) but not glucose (+0.012 ± 0.071 mmol/liter, P = 0.86). Similarly, fasting non-HDL-C (Fig. 3B), apoB (Fig. 3C), and the apoB to apoAI ratio (Fig. 3D) were all significantly increased in subjects consuming fructose (non-HDL-C: +0.29 ± 0.066 mmol/liter, P = 0.0081; apoB: +0.093 ± 0.022 g/liter, P = 0.0005; apoB to apoAI ratio: +14.6 ± 3.8%, P = 0.0006) and HFCS (non-HDL-C: +0.55 ± 0.14 mmol/liter, P < 0.0001; apoB: +0.12 ± 0.031 g/liter, P < 0.0001; apoB to apoAI ratio: +19.5 ± 4.4%, P < 0.0001) compared with baseline but not in subjects consuming glucose (non-HDL-C: +0.055 ± 0.080 mmol/liter, P = 0.49; apoB: +0.0097 ± 0.019 g/liter, P = 0.90; apoB to apoAI ratio: +1.9 ± 2.5%, P = 0.81).

TABLE 2: Primary outcomes during consumption of complex carbohydrates at 0 wk and during consumption of glucose-, fructose-, or HFCS-sweetened beverages at 2 wk

Primary outcomes	Glucose		Fructose		HFCS		Effects	P value
	0 wk	2 wk	0 wk	2 wk	0 wk	2 wk		
24-h TG AUC (mmol/liter per 24 h)[a]	5.6 ± 1.1	3.6 ± 1.3[b]	2.9 ± 1.5	7.6 ± 1.9*[c]	3.8 ± 1.4	5.5 ± 1.7**[c]	Sugar	0.0058

TABLE 2: *Cont.*

Primary outcomes	Glucose		Fructose		HFCS		Effects	P value
	0 wk	2 wk	0 wk	2 wk	0 wk	2 wk		
							Gender	0.13
							BMI	0.0033
Late-evening TG (mmol/liter)[d]	1.3 ± 0.2	1.5 ± 0.2^{b}	1.2 ± 0.1	$1.8 \pm 0.2^{***,c}$	1.3 ± 0.2	$1.8 \pm 0.2^{***,b,c}$	Sugar	0.016
							Gender	0.40
							BMI	0.015
Fasting TG (mmol/liter)[e]	1.2 ± 0.1	$1.4 \pm 0.2^{*}$	1.2 ± 0.1	1.3 ± 0.1	1.3 ± 0.2	1.4 ± 0.1	Sugar	0.54
							Gender	0.035
							BMI	0.94
Fasting LDL-C (mmol/liter)[a]	2.6 ± 0.2	2.6 ± 0.2^{b}	2.1 ± 0.2	$2.4 \pm 0.2^{*,b,c}$	2.3 ± 0.2	$2.7 \pm 0.2^{***,c}$	Sugar	0.0098
							Gender	0.057
							BMI	0.40
Fasting non-HDL-C (mmol/liter)[a]	3.2 ± 0.2	3.3 ± 0.2^{b}	2.7 ± 0.2	$3.0 \pm 0.2^{*,b,c}$	2.9 ± 0.2	$3.4 \pm 0.2^{***,c}$	Sugar	0.0077
							Gender	0.017
							BMI	0.48
Fasting apoB (g/liter)[a]	0.82 ± 0.06	0.83 ± 0.06^{b}	0.65 ± 0.04	$0.74 \pm 0.05^{****,c}$	0.73 ± 0.05	$0.85 \pm 0.06^{***,c}$	Sugar	0.0051
							Gender	0.027
							BMI	0.47
apoB to apoAI ratio[e]	0.70 ± 0.06	0.70 ± 0.05^{b}	0.54 ± 0.04	$0.63 \pm 0.05^{*,c}$	0.60 ± 0.06	$0.71 \pm 0.07^{***,c}$	Sugar	0.0031
							Gender	0.34
							BMI	0.61

P > 0.05 for differences among groups at baseline for all outcomes. Mean ± sem. [a]PROC MIXED two-factor (sugar, gender) analysis with adjustment for BMI, ΔBW (2 wk to 0 wk), and outcomeB on absolute Δ (2 wk vs. 0 wk). [b]Δ (2 wk vs. 0 wk) significantly different from [c]Δ (2 wk vs. 0 wk), Tukey's multiple comparison test. [d]PROC MIXED two-factor (sugar, gender) analysis with adjustment for BMI on absolute Δ (2 wk vs. 0 wk). [e]PROC MIXED two-factor (sugar, gender) analysis with adjustment for BMI and Δ BW (2 wk to 0 wk) on absolute Δ (2 wk vs. 0 wk). *P < 0.01, **P < 0.05, ***P < 0.0001, ****P < 0.001, LS means of Δ different from zero.

TABLE 3: Secondary outcomes during consumption of complex carbohydrates at 0 wk and during consumption of glucose-, fructose-, or HFCS-sweetened beverages at 2 wk

Secondary outcomes	Glucose		Fructose		HFCS		Effects	P value
	0 wk	2 wk	0 wk	2 wk	0 wk	2 wk		
Body weight (kg)[a]	76.8 ± 3.5	77.2 ± 3.7	76.8 ± 2.6	76.7 ± 2.6	74.3 ± 3.7	74.7 ± 3.7	Sugar	0.32
							Gender	0.62
							BMI	0.50
Fasting HDL (mmol/ liter)[b]	1.2 ± 0.1	1.2 ± 0.1	1.2 ± 0.1	1.1 ± 0.1	1.2 ± 0.1	1.2 ± 0.1	Sugar	0.92
							Gender	0.37
							BMI	0.22
PP LDL (mmol/ liter)[b]	2.5 ± 0.2	2.6 ± 0.2[c]	2.0 ± 0.2	2.3 ± 0.2[*,c,d]	2.1 ± 0.2	2.7 ± 0.2[**,d]	Sugar	0.0033
							Gender	0.010
							BMI	0.54
PP non-HDL-C (mmol/ liter)[b]	3.0 ± 0.2	3.2 ± 0.2[c]	2.5 ± 0.2	3.0 ± 0.2[**,c,d]	2.6 ± 0.2	3.4 ± 0.2[**,d]	Sugar	0.0012
							Gender	0.017
							BMI	0.27
PP apoB (g/liter)[b]	0.78 ± 0.05	0.83 ± 0.05[c]	0.62 ± 0.04	0.73 ± 0.05[***,c,b]	0.68 ± 0.05	0.84 ± 0.06[**,d]	Sugar	0.031
							Gender	0.10
							BMI	0.56
PP RLP-C (mmol/ liter)[a]	0.17 ± 0.02	0.19 ± 0.02[c]	0.16 ± 0.02	0.23 ± 0.03[**,b]	0.15 ± 0.02	0.21 ± 0.02[***,c,b]	Sugar	0.035
							Gender	0.37
							BMI	0.034
PP RLP-TG (mmol/ liter)[e]	0.34 ± 0.07	0.44 ± 0.06	0.35 ± 0.06	0.58 ± 0.08[***]	0.33 ± 0.06	0.54 ± 0.09[***]	Sugar	0.088
							Gender	0.20
							BMI	0.012

TABLE 3: *Cont.*

Secondary outcomes	Glucose		Fructose		HFCS		Effects	P value
	0 wk	2 wk	0 wk	2 wk	0 wk	2 wk		
Fasting sdLDL-C (mmol/liter)f	0.65 ± 0.08	0.77 ± 0.10****	0.47 ± 0.04	0.59 ± 0.06***	0.61 ± 0.08	0.78 ± 0.09**	Sugar	0.37
							Gender	0.0019
							BMI	0.11
PP sdLDL-C (mmol/liter)f	0.65 ± 0.08	0.79 ± 0.10*,c	0.48 ± 0.04	0.64 ± 0.07**,c,d	0.60 ± 0.08	0.86 ± 0.10**,d	Sugar	0.019
Gender	<0.0001							
BMI	0.0125							

$P > 0.05$ for differences among groups at baseline for all outcomes. Mean ± sem. PP, Postprandial. aPROC MIXED two-factor (sugar, gender) analysis with adjustment for BMI on absolute Δ (2 wk vs. 0 wk). bPROC MIXED two-factor (sugar, gender) analysis with adjustment for BMI, ΔBW (2 wk to 0 wk), and outcomeB on absolute Δ (2 wk vs. 0 wk). cΔ (2 wk vs. 0 wk) significantly different from dΔ (2 wk vs. 0 wk), Tukey's multiple comparison test. ePROC MIXED two-factor (sugar, gender) analysis with adjustment for BMI and outcomeB on absolute Δ (2 wk vs. 0 wk). fPROC MIXED two-factor (sugar, gender) analysis with adjustment for BMI and ΔBW (2 wk to 0 wk) on absolute Δ (2 wk vs. 0 wk). *$P < 0.01$, **$P < 0.0001$, ***$P < 0.001$, ****$P < 0.05$, LS means of Δ different from zero.

3.2.2 PRIMARY OUTCOMES: COMPARING GLUCOSE, FRUCTOSE, AND HFCS CONSUMPTION

The effects of the three sugars were significantly different (PROC MIXED two factor analysis with adjustment for BMI, ΔBW and outcomeB) for all primary outcomes except fasting TG (see effects of sugar P values in Table 2). The effects of HFCS compared with fructose consumption on all primary outcomes were not significantly different ($P > 0.05$, Tukey's). The increases in 24-h TG AUC ($P = 0.0068$), late evening TG peaks ($P = 0.015$), fasting apoB ($P = 0.037$), and the apoB to apoA1 ratio ($P = 0.028$) were larger after fructose consumption compared with glucose consumption. The increases in 24-h TG AUC ($P = 0.034$), fasting LDL ($P = 0.0083$), non-HDL-C ($P = 0.0055$), apoB ($P = 0.0056$), and apoB to apoAI ratio ($P = 0.0034$) were larger after HFCS consumption than glucose consumption.

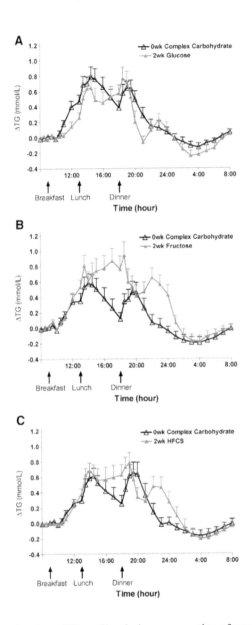

FIGURE 1: Twenty-four-hour TG profiles during consumption of complex carbohydrate and during consumption of sugar-sweetened beverages. The change of 24-h TG concentrations over fasting concentrations during consumption of energy-balanced baseline diet containing 55% E complex carbohydrate at 0 wk and during consumption of energy-balanced intervention diet containing 30% E complex carbohydrate and 25% E glucose (A), fructose (B), or HFCS (C) at 2 wk (n = 16/group). Data are mean ± sem.

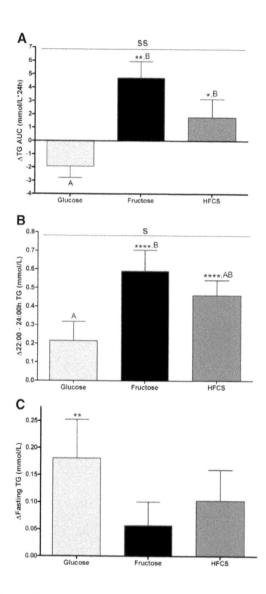

FIGURE 2: Effects of sugar-sweetened beverage consumption on TG concentrations. The change in 24-h TG AUC (A), late-night to late-evening TG (B), and fasting TG concentrations (C) compared with baseline after consuming 25% of energy requirements as glucose-, fructose-, or HFCS-sweetened beverages for 2 wk is shown. S, $P < 0.05$; SS, $P < 0.01$, effect of sugar; two-factor (sugar, gender) PROC MIXED analysis on Δ with adjustment for BMI (B), ΔBW (C), and outcome at baseline (A). *, $P < 0.05$, **, $P < 0.01$, ****, $P < 0.0001$, LS means different from zero. A, Δ different from B; Δ, Tukey's (n = 16/ group). Data are mean ± sem.

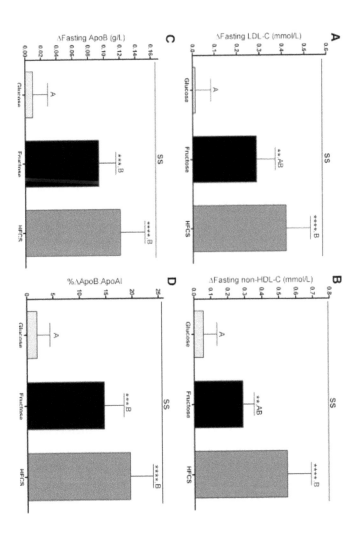

FIGURE 3: Effects of sugar-sweetened beverage consumption on risk factors for cardiovascular disease. The change in fasting LDL-C (A), non-HDL-C (B), apoB concentrations (C), and apoB to apoA1 ratio (D) after consuming 25% of energy requirements as glucose-, fructose-, or HFCS-sweetened beverages for 2 wk. ss, P < 0.01, effect of sugar; two-factor (sugar, gender) PROC MIXED analysis on Δ with adjustment for BMI, ΔBW (D), and outcome at baseline (A–C). **, P < 0.01, ***, P < 0.001, ****, P < 0.0001, LS means different from zero. A, Δ different from B, Δ, Tukey's (n = 16/group). Data are mean ± sem.

With regard to BMI, although the statistical results presented in Table 1 and Figs. 2 and 3 include adjustment for BMI, Online Supplemental Fig. 1, A–F, presents the changes of the primary outcomes with subjects grouped as normal weight (BMI <25 kg/m^2) or overweight/obese (BMI >25 kg/m^2). The effect of BMI status was significant for the change of the 24-h TG AUC (P = 0.016) and the late-evening TG peaks (P = 0.019) but not for the fasting TG (P = 0.55, data not shown), LDL-C (P = 0.30), non-HDL-C (P = 0.93), apoB (P = 0.62), and apoB to apoAI ratio (P = 0.51). Normal weight and overweight/obese subjects consuming HFCS had comparable absolute (Supplemental Fig. 1) and percent increases of late-evening TG (BMI <25 kg/m^2: +46 ± 11%; BMI >25 kg/m^2: +31 ± 6%), fasting LDL-C (BMI <25 kg/m^2: +22 ± 1%; BMI >25 kg/m^2: +28 ± 1%), non-HDL-C (BMI <25 kg/m^2: +36 ± 19%, BMI >25 kg/m^2: +17 ± 7), apoB (BMI < 25 kg/m^2: +17 ± 6%; BMI >25 kg/m^2: +20 ± 8%), and the apoB to apoAI ratio (BMI <25 kg/m^2: +22 ± 7%; BMI >25 kg/m^2: +20 ± 9%).

3.2.3 SECONDARY OUTCOMES: COMPARING GLUCOSE, FRUCTOSE, AND HFCS WITH COMPLEX CARBOHYDRATE CONSUMPTION

Table 3 presents the secondary outcomes during consumption of complex carbohydrate at baseline and at the end of the 2-wk sugar interventions. Fasting HDL concentrations were unaffected by consumption of the three sugar-sweetened beverages. Similar to the responses in the fasting state, subjects consuming fructose and HFCS had increased postprandial concentrations of LDL-C, non-HDL-C, and apoB compared with baseline, whereas subjects consuming glucose did not. Fructose and HFCS consumption increased postprandial concentrations of RLP-C and RLP-TG compared with baseline, whereas consumption of glucose did not. Consumption of all three sugars increased fasting and postprandial sdLDL-C compared with baseline.

3.2.4 SECONDARY OUTCOMES: COMPARING GLUCOSE, FRUCTOSE, AND HFCS CONSUMPTION

The effects of the three sugars were significantly different (PROC MIXED two factor analysis with adjustment for BMI, ΔBW, and outcomeB) for postprandial LDL, non-HDL-C, apoB, RLP-C, and sdLDL-C (see effects of sugar P values in Table 3). The effects of HFCS compared with fructose consumption on all secondary outcomes were not significantly different (P > 0.05, Tukey's). The increases in postprandial RLP-C were larger during consumption of fructose compared with glucose (P = 0.044), and HFCS consumption caused larger increases in postprandial LDL (P = 0.0024), non-HDL-C (P = 0.0007), apoB (P = 0.025), and sdLDL-C (P = 0.014) (Tukey's) than glucose consumption.

For glucose, insulin, and HOMA-IR, the 24-h glucose and insulin profiles during baseline (0 wk) and at the end of the 2-wk intervention are presented in Online Supplemental Figs. 2, A–C, and 3, A–C, respectively. Compared with baseline, the 24-h glucose and insulin 24-h AUC and the postmeal insulin peaks were significantly increased in subjects consuming glucose, significantly decreased in subjects consuming fructose, and were unchanged in subjects consuming HFCS (Online Supplemental Table 1). Postmeal glucose peaks were increased in subjects consuming glucose and HFCS. Fasting glucose concentrations were significantly decreased in subjects consuming glucose, whereas fasting insulin concentrations and HOMA-IR did not change significantly in any group.

3.2.5 GENDER

Although there were no significant sugar-gender interactions for any of the primary or secondary outcomes, men exhibited larger increases of fasting TG, non-HDL-C, apoB, and sdLDL-C concentrations and postprandial LDL, non-HDL-C, and sdLDL-C concentrations in response to sugar consumption than women (see effects of gender P values in Tables 2 and 3).

However, postprandial TG responses, as assessed by the 24-h TG AUC, late-evening TG peaks, postprandial apoB, and RLP-TG concentrations, were not different between genders. The subjects consuming glucose exhibited the most divergent gender responses, particularly in sdLDL-C. Fasting and postprandial sdLDL-C levels were increased compared with baseline by $+0.22 \pm 0.07$ mmol/liter (P = 0.0001) and $+0.24 \pm 0.05$ mmol/liter (P < 0.0001), respectively, in men after glucose consumption but were unchanged in women (fasting sdLDL-C: -0.004 ± 0.02 mmol/liter, P = 0.61; postprandial sdLDL-C: $+0.006 \pm 0.019$ mmol/liter, P = 0.69).

3.3 DISCUSSION

The current study provides evidence that postprandial TG and fasting and postprandial concentrations of LDL, non-HDL-C, apoB, and the apoB to apoAI ratio, established risk factors for coronary heart disease (25), are significantly increased in response to 2 wk consumption of 25% of E as fructose and HFCS, but not glucose, in younger, normal-weight, and overweight subjects. In contrast and as was observed in older subjects (15), fasting TG concentrations were increased in subjects consuming glucose but not in those consuming fructose-containing sugars. The differential effects of fructose and glucose consumption on fasting and postprandial TG responses in subjects from both studies suggest that fasting TG concentrations are not a reliable indicator of the adverse changes in postprandial TG and other lipid/lipoprotein risk factors induced by fructose consumption. There is growing evidence linking increases of postprandial TG concentrations with proatherogenic conditions (26–28). It is important to note that for both the current and previous study (15), the differential effects of fructose and HFCS compared with complex carbohydrate on the 24-h TG profile were most marked in the late evening, approximately 4 and 6 h after dinner. Studies investigating the relationship between this late-evening peak and proatherogenic changes would be of interest, as would investigations into the sources of the TG that contributes to these peaks (de novo lipogenesis, diet, or fatty acids derived from adipose lipolysis).

To our knowledge this is the first study to directly compare the effects of sustained consumption of HFCS with 100% fructose and glucose-sweetened beverages. This comparison is important because it would seem likely that the effects of HFCS-sweetened beverages on circulating lipids and lipoproteins would be less than those of pure fructose-sweetened beverage because they contain 45% less fructose. And indeed, the postprandial TG and RLP responses exhibited the expected pattern based on the fructose content of the sugars, with increases being greatest in subjects who consumed 145 ± 4 g fructose per day from beverages, lowest in subjects who consumed 144 ± 5 g glucose per day and 0 g fructose per day from beverages and intermediate in subjects who consumed HFCS-sweetened beverages providing 64 ± 2 g glucose per day and 79 ± 3 g fructose per day. However, the changes of fasting and postprandial concentrations of LDL, non-HDL-C, apoB, and the apoB to apoAI ratio in subjects consuming HFCS were significantly larger compared with subjects consuming glucose and tended to be higher compared with subjects consuming pure fructose. More studies are needed to confirm this unexpected pattern and to determine whether it is a result of a synergistic effect of consuming fructose and glucose in combination. Additional studies are also needed to determine whether the substantial increases, seen after just 2 wk, are further aggravated with longer-term consumption of HFCS-sweetened beverages.

Compared with baseline, postmeal glucose and insulin responses (indexed as 24 h AUC and postmeal peaks) were mainly increased during glucose consumption, decreased during fructose consumption, and unchanged during HFCS consumption. This pattern is expected and further supports our data indicating that the adverse effects associated with chronic consumption of sugar-sweetened beverages result from the specific effects of fructose (29), rather than from increased circulating glucose and insulin excursions (i.e. glycemic index) (30–32). Although consumption of fructose increased fasting glucose and insulin concentrations in 2 wk and decreased insulin sensitivity by 17% in 10 wk (15), in the current study, HOMA-IR was unchanged after 2 wk consumption of fructose, HFCS, or glucose. This may be related to the subjects in the current study being younger and leaner (28 ± 7 yr; 25.5 ± 4.0 kg/m^2) than the subjects in the previous study (54 ± 8 yr; 29.1 ± 2.9 kg/m^2). In a study by Le et al. (33), inclusion of fructose with an energy-balanced diet for 4 wk in young, normal-weight men (24.7 ± 1.3 yr; -22 kg/m^2) increased fasting glucose levels, but other indices of insulin sensitivity were unaffected. However, it was recently reported that consumption of fructose or glucose (150 g/d) for 4 wk lowered insulin sensitivity and increased HOMA-IR in subjects of similar age and BMI (31 ± 9 yr; 25.9 ± 2.2 kg/m^2) (34).

As would be expected based on the evidence that both increasing age and postmenopausal status result in augmented postprandial lipid responses in women (35), more significant gender differences in lipid outcomes were observed in these younger subjects in the current study than in the older subjects previously studied (15). With the exception of postprandial TG, apoB, and RLP-C and RLP-TG, younger men exhibited larger lipoprotein responses after 2 wk of sugar consumption than younger women. The comparable responses in postprandial TG and apoB concentrations and the significantly different fasting TG and apoB responses between the genders suggest that rates of very low-density lipoprotein secretion may be similar between men and women, whereas rates of very low-density lipoprotein clearance are different. This is supported by kinetic studies, which demonstrate that women have higher TG-rich lipoprotein and LDL-apoB fractional catabolic rates than men, whereas production rates are comparable (36, 37).

The greater effect of glucose consumption on sdLDL-C levels in younger men compared with younger women represents the most marked difference between the current and our previous lipid results, which showed older men and women were comparably nonresponsive to consumption of glucose (15). The increase of fasting sdLDL-C concentrations compared with baseline in younger men consuming glucose was unexpected because they did not exhibit increases in fasting LDL and apoB concentrations.

The added sugar component of the typical U.S. diet consists of nearly equal amounts of HFCS and sucrose (38); therefore, it is a limitation of this study that we did not also investigate the effects of sucrose consumption. However, we expect that the effects of sucrose would be comparable with those of HFCS because its composition (50% glucose/50% fructose) is very similar to the composition of the HFCS used for this study (45% glucose/55% fructose). This is supported by results from a crossover study in which subjects consumed standardized diets containing 5, 18, or 33% of energy as sucrose, each for 6 wk. Compared with the 5% sucrose diet,

LDL concentrations increased by 17% on the 18% sucrose diet and by 22% on the 33% sucrose diet (39).

Self-reported intake data suggest that 13% of the U.S. population consumes 25% or more of energy from added sugar (40). Importantly, the current results provide evidence that sugar consumption at this level increases risk factors for cardiovascular disease within 2 wk in young adults, thus providing direct experimental support for the epidemiological evidence linking sugar consumption with dyslipidemia (1–5) and cardiovascular disease (13). The results contradict the conclusions from recent reviews that sugar intakes as high as 25–50% of energy have no adverse long-term effects with respect to components of the metabolic syndrome (16) and that fructose consumption up to 140 g/d does not result in a biologically relevant increase of fasting or postprandial TG in healthy, normal-weight (17) or overweight or obese (18) humans. Additionally they provide evidence that the maximal upper limit of 25% of total energy requirements from added sugar, suggested by the Dietary Guidelines for Americans 2010 (19), may need to be reevaluated.

REFERENCES

1. Aeberli I , Zimmermann MB , Molinari L , Lehmann R , l'Allemand D , Spinas GA, Berneis K 2007 Fructose intake is a predictor of LDL particle size in overweight schoolchildren. Am J Clin Nutr 86:1174–1178
2. Bortsov AV , Liese AD , Bell RA , Dabelea D , D'Agostino RB , Hamman RF , Klingensmith GJ , Lawrence JM , Maahs DM , McKeown R , Marcovina SM , Thomas J , Williams DE , Mayer-Davis EJ 20 January 2011 Sugar-sweetened and diet beverage consumption is associated with cardiovascular risk factor profile in youth with type 1 diabetes. Acta Diabetol 10.1007/s00592-010-0246-9
3. Duffey KJ , Gordon-Larsen P , Steffen LM , Jacobs DR , Popkin BM 2010 Drinking caloric beverages increases the risk of adverse cardiometabolic outcomes in the Coronary Artery Risk Development in Young Adults (CARDIA) Study. Am J Clin Nutr 92:954–959
4. Welsh JA , Sharma A , Abramson JL , Vaccarino V , Gillespie C , Vos MB 2010 Caloric sweetener consumption and dyslipidemia among U.S. adults. JAMA 303:1490–1497
5. Welsh JA , Sharma A , Cunningham SA , Vos MB 2011 Consumption of added sugars and indicators of cardiovascular disease risk among U.S. adolescents. Circulation 123:249–257

6. Bremer AA , Auinger P , Byrd RS 6 September 2009 Sugar-sweetened beverage intake trends in U.S. adolescents and their association with insulin resistance-related parameters. J Nutr Metab 10.1155/2010/196476

7. Yoshida M , McKeown NM , Rogers G , Meigs JB , Saltzman E , D'Agostino R , Jacques PF 2007 Surrogate markers of insulin resistance are associated with consumption of sugar-sweetened drinks and fruit juice in middle and older-aged adults. J Nutr 137:2121–2127

8. Assy N , Nasser G , Kamayse I , Nseir W , Beniashvili Z , Djibre A , Grosovski M 2008 Soft drink consumption linked with fatty liver in the absence of traditional risk factors. Can J Gastroenterol 22:811–816

9. Ouyang X , Cirillo P , Sautin Y , McCall S , Bruchette JL , Diehl AM , Johnson RJ , Abdelmalek MF 2008 Fructose consumption as a risk factor for non-alcoholic fatty liver disease. J Hepatol 48:993–999

10. Montonen J , Järvinen R , Knekt P , Heliövaara M , Reunanen A 2007 Consumption of sweetened beverages and intakes of fructose and glucose predict type 2 diabetes occurrence. J Nutr 137:1447–1454

11. Palmer JR , Boggs DA , Krishnan S , Hu FB , Singer M , Rosenberg L 2008 Sugar-sweetened beverages and incidence of type 2 diabetes mellitus in African American women. Arch Intern Med 168:1487–1492

12. Schulze MB , Manson JE , Ludwig DS , Colditz GA , Stampfer MJ , Willett WC , Hu FB 2004 Sugar-sweetened beverages, weight gain, and incidence of type 2 diabetes in young and middle-aged women. JAMA 292:927–934

13. Fung TT , Malik V , Rexrode KM , Manson JE , Willett WC , Hu FB 2009 Sweetened beverage consumption and risk of coronary heart disease in women. Am J Clin Nutr 89:1037–1042

14. Dhingra R , Sullivan L , Jacques PF , Wang TJ , Fox CS , Meigs JB , D'Agostino RB , Gaziano JM , Vasan RS 2007 Soft drink consumption and risk of developing cardiometabolic risk factors and the metabolic syndrome in middle-aged adults in the community. Circulation 116:480–488

15. Stanhope KL , Schwarz JM , Keim NL , Griffen SC , Bremer AA , Graham JL , Hatcher B , Cox CL , Dyachenko A , Zhang W , McGahan JP , Seibert A , Krauss RM, Chiu S , Schaefer EJ , Ai M , Otokozawa S , Nakajima K , Nakano T , Beysen C, Hellerstein MK , Berglund L , Havel PJ 2009 Consuming fructose-sweetened, not glucose-sweetened, beverages increases visceral adiposity and lipids and decreases insulin sensitivity in overweight/obese humans. J Clin Invest 119:1322–1334

16. Ruxton CH , Gardner EJ , McNulty HM 2010 Is sugar consumption detrimental to health? A review of the evidence, 1995–2006. Crit Rev Food Sci Nutr 50:1–19

17. Dolan LC , Potter SM , Burdock GA 2010 Evidence-based review on the effect of normal dietary consumption of fructose on development of hyperlipidemia and obesity in healthy, normal weight individuals. Crit Rev Food Sci Nutr 50:53–84

18. Dolan LC , Potter SM , Burdock GA 2010 Evidence-based review on the effect of normal dietary consumption of fructose on blood lipids and body weight of overweight and obese individuals. Crit Rev Food Sci Nutr 50:889–918

19. DGAC 2010 Report of the Dietary Guidelines Advisory Committee (DGAC) on the dietary guidelines for Americans, 2010. http://www.cnpp.usda.gov/dgas2010-dgacreport.htm

20. Johnson RK , Appel LJ , Brands M , Howard BV , Lefevre M , Lustig RH , Sacks F, Steffen LM , Wylie-Rosett J 2009 Dietary sugars intake and cardiovascular health: a scientific statement from the American Heart Association. Circulation 120:1011–1020

21. Mifflin MD , St Jeor ST , Hill LA , Scott BJ , Daugherty SA , Koh YO 1990 A new predictive equation for resting energy expenditure in healthy individuals. Am J Clin Nutr 51:241–247

22. Sakaue T , Hirano T , Yoshino G , Sakai K , Takeuchi H , Adachi M 2000 Reactions of direct LDL-cholesterol assays with pure LDL fraction and IDL: comparison of three homogeneous methods. Clin Chim Acta 295:97–106

23. Ito Y , Fujimura M , Ohta M , Hirano T 2011 Development of a homogeneous assay for measurement of small dense LDL cholesterol. Clin Chem 57:57–65

24. Nakajima K , Saito T , Tamura A , Suzuki M , Nakano T , Adachi M , Tanaka A , Tada N , Nakamura H , Murase T 1994 A new approach for the detection of type III hyperlipoproteinemia by RLP-cholesterol assay. J Atheroscler Thromb 1:30–36

25. Di Angelantonio E , Sarwar N , Perry P , Kaptoge S , Ray KK , Thompson A , Wood AM , Lewington S , Sattar N , Packard CJ , Collins R , Thompson SG , Danesh J 2009 Major lipids, apolipoproteins, and risk of vascular disease. JAMA 302:1993–2000

26. Bansal S , Buring JE , Rifai N , Mora S , Sacks FM , Ridker PM 2007 Fasting compared with nonfasting triglycerides and risk of cardiovascular events in women. JAMA 298:309–316

27. Karpe F 1999 Postprandial lipoprotein metabolism and atherosclerosis. J Intern Med 246:341–355

28. Nordestgaard BG , Benn M , Schnohr P , Tybjaerg-Hansen A 2007 Nonfasting tri-glycerides and risk of myocardial infarction, ischemic heart disease, and death in men and women. JAMA 298:299–308

29. Stanhope KL , Griffen SC , Bremer AA , Vink RG , Schaefer EJ , Nakajima K , Schwarz JM , Beysen C , Berglund L , Keim NL , Havel PJ 2011 Metabolic re-sponses to prolonged consumption of glucose- and fructose-sweetened beverages are not associated with postprandial or 24-h glucose and insulin excursions. Am J Clin Nutr 94:112–119

30. Ding EL , Malik VS 2008 Convergence of obesity and high glycemic diet on com-pounding diabetes and cardiovascular risks in modernizing China: an emerging pub-lic health dilemma. Global Health 4:4

31. Hu FB , Malik VS 2010 Sugar-sweetened beverages and risk of obesity and type 2 diabetes: epidemiologic evidence. Physiol Behav 100:47–54

32. Schernhammer ES , Hu FB , Giovannucci E , Michaud DS , Colditz GA , Stampfer MJ , Fuchs CS 2005 Sugar-sweetened soft drink consumption and risk of pancreatic cancer in two prospective cohorts. Cancer Epidemiol Biomarkers Prev 14:2098–2105

33. Lê KA , Faeh D , Stettler R , Ith M , Kreis R , Vermathen P , Boesch C , Ravussin E, Tappy L 2006 A 4-wk high-fructose diet alters lipid metabolism without affecting insulin sensitivity or ectopic lipids in healthy humans. Am J Clin Nutr 84:1374–1379

34. Silbernagel G , Machann J , Unmuth S , Schick F , Stefan N , Häring HU , Fritsche A 2011 Effects of 4-week very-high-fructose/glucose diets on insulin sensitivity, visceral fat and intrahepatic lipids: an exploratory trial. Br J Nutr 106:79–86

35. Jackson KG , Abraham EC , Smith AM , Murray P , O'Malley B , Williams CM , Minihane AM 2010 Impact of age and menopausal status on the postprandial triacylglycerol response in healthy women. Atherosclerosis 208:246–252

36. Matthan NR , Jalbert SM , Barrett PH , Dolnikowski GG , Schaefer EJ , Lichtenstein AH 2008 Gender-specific differences in the kinetics of nonfasting TRL, IDL, and LDL apolipoprotein B-100 in men and premenopausal women. Arterioscler Thromb Vasc Biol 28:1838–1843

37. Watts GF , Moroz P , Barrett PH 2000 Kinetics of very-low-density lipoprotein apolipoprotein B-100 in normolipidemic subjects: pooled analysis of stable-isotope studies. Metabolism 49:1204–1210

38. Wells FW , Buzby JC 2008 Dietary assessment of major trends in U.S. food consumption, 1970–2005. Economic Information Bulletin No. (EIB-33). Washington, DC: Department of Agriculture

39. Reiser S , Bickard MC , Hallfrisch J , Michaelis OE , Prather ES 1981 Blood lipids and their distribution in lipoproteins in hyperinsulinemic subjects fed three different levels of sucrose. J Nutr 111:1045–1057

40. Marriott BP , Olsho L , Hadden L , Connor P 2010 Intake of added sugars and selected nutrients in the United States, National Health and Nutrition Examination Survey (NHANES), 2003–2006. Crit Rev Food Sci Nutr 50:228–258

CHAPTER 4

Treatment of Metabolic Syndrome by Combination of Physical Activity and Diet Needs an Optimal Protein Intake: A Randomized Controlled Trial

FRÉDÉRIC DUTHEIL, GÉRARD LAC, DANIEL COURTEIX,
ERIC DORÉ, ROBERT CHAPIER, LAURENCE ROSZYK,
VINCENT SAPIN, AND BRUNO LESOURD

4.1 INTRODUCTION

The International Diabetes Federation (IDF) defines the metabolic syndrome (MS) as the co-occurrence of any three of the five following abnormalities : abdominal obesity (waist circumference > 94 cm in men and > 80 in women), dyslipidemia (triglyceridemia > 1.5 mmol/l, HDL cholesterol < 0.4 g/l in men and < 0.5 g/l in women), blood pressure (BP) > 130/85 and/or medical treatment, and fasting glycemia > 5.55 mmol/l and/or medical treatment [1]. MS is associated with an increased risk of cardiovascular diseases [2] and prevalence of type 2 diabetes [3]. In developed countries, its increasing prevalence is mainly linked to obesity and age [4].

The most efficient strategy to counteract MS is a significant reduction in caloric intake associated with an increase in physical activity (PA). Such programmes aim primarily to reduce overweight, the most visible manifestation of MS, but the challenge is to reduce the fat mass without affecting lean body mass, especially in senior, for whom a progressive loss of muscle mass and strength is a natural phenomenon [5], even in those who are healthy and physically active [6]. In addition, the recovery of skeletal muscle mass in ageing people is impaired after a catabolic state [7,8]. Physical exercise and an adequate protein intake are of prime importance in preventing muscle loss. However, there is no consensus on the adequate level of protein intake in the case of senior patients undergoing a combined treatment of caloric restriction and physical activity (PA) for MS. In these patients, age, exercise and energy restriction increase protein requirement.

The recommended dietary protein allowance (RDA) for the general population has been set at 0.8 g/kg/day [9,10]. RDA is defined as the average daily dietary intake level that is sufficient to meet the nutrient requirements of nearly all healthy individuals. RDA corresponds to the mean lower threshold intake (LTI) of a panel of healthy people plus two standard deviations, including 97.5% of the population, and is calculated as 1.3 LTI day [9].

Some guidelines recommend increasing RDA to 1.0–1.3 g/kg/day in senior [11]. PA increases the need for proteins whatever the age of the subject [9,12,13], and this specificity must be taken into account in senior people [14,15].

Total energy intake has a protein sparing effect [16-18]. Conversely insufficient energy intake will increase the protein needed to compensate for the energy deficit. As skeletal muscle is the main storage site of body proteins and amino acids, this will lead to an undesirable reduction of muscle mass [19]. Excessive protein intake is of no value, in particular because it will over-exert the kidney [20] and increase the end products of protein metabolism (urea and uric acid). It will also increase the intake of undesirable saturated fatty acids via proteins of animal origin [21]. The precautionary principle is to bear this factor in mind in senior patients, since age-related renal insufficiency is common [22], especially in people with elevated BP [23] and/or dyslipidemia [24], which is often the case in subjects with MS.

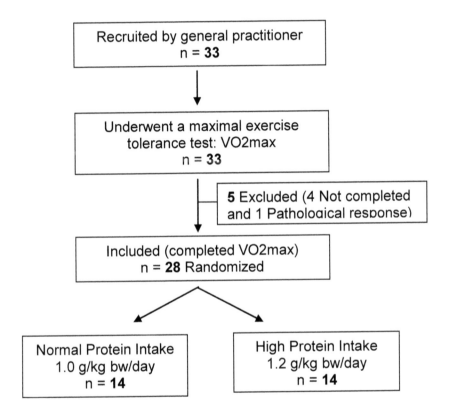

FIGURE 1: Flow chart of the study design.

The challenge for the prescriber is to give neither too much nor too little protein, in order to preserve the muscle mass without inducing harmful effects on the kidney in older subjects with MS.

Our aim in the present study was to assess the minimal need for proteins in a population of senior MS subjects. There are a limited number of tools to assess the appropriate level of protein intake. One way is to control preservation of muscle mass over a long period, but this can only be done in animal studies for ethical reasons. Nitrogen balance studies are probably the gold standard, but they are rather cumbersome to perform. Monitoring the levels of albumin, the blood marker of protein metabolism homeostasis, seems to be the most convenient index and was chosen for this study. Moreover, albumin levels are closely linked to morbidity, and represent a large consensus to assess nutritional status [25,26]. We decided to determine protein LTI by recording albumin levels in older subjects with MS participating in a weight reduction programme including exercise and energy restriction. The programme comprised two parts: a three-week residential programme during which subjects stayed in a medical establishment on a controlled diet with regular PA, and a six-month follow-up at home.

4.2 SUBJECTS AND METHODS

4.2.1 PARTICIPANTS

We needed to recruit between 25 and 30 volunteers, of both sexes, aged from 50 to 70 years, presenting the characteristics of the MS as defined by the IDF criteria in 2005 [1]. Potential participants underwent a comprehensive medical screening procedure. Volunteers were eligible for inclusion in the study if they had a sedentary lifestyle, and stable body weight over the previous year (i.e., had not fluctuated more than 2 kg), and if their medical treatment had remained the same during the 6 months before recruitment. Major criteria for exclusion were the presence of cardiovascular, hepatic, renal or endocrine diseases, the use of medications that affect body weight, restricted diet in the past year, insufficient motivation as assessed by interview, and inability to complete a maximal exercise tolerance test (VO2max).

Of the 33 participants with MS recruited by their general practitioner, 4 were unable to complete a maximal exercise tolerance test (VO2max) and 1 had a pathological response. Twenty eight volunteers, 19 men, 9 women, aged 61.8±6.5 years, were included. All were Caucasians. They were randomly assigned to two groups of different PI, normal and high. They all completed the study (Figure 1).

The study was approved by the local "Committee for the Protection of the Person for Research in Biology" (CPPRB). All participants gave written informed consent. They were informed that the study would be comparing diets with two different protein intakes and that they would be assigned a diet at random. Random assignments to one of two different protein intake groups were computer-generated after subjects were considered eligible to take part.

4.2.2 STUDY OUTCOMES

The primary outcome was the change from baseline in albumin levels. Secondary outcomes included other markers of protein intake such as body composition, in particular lean mass, creatinine levels, renal clearance and pro-inflammatory factors such as C-Reactive Protein (CRP) and orosomucoid.

4.2.3 STUDY DESIGN

In this 26-week study, participants were randomly assigned, with stratification according to sex and weight, to one of two groups: a normal PI group (NPI) with intake of 1.0 g/kg/d and a high PI group (HPI) with 1.2 g/kg/d.

The study design is shown in Figure 2. The study comprised three chronological stages: Day 0 (D0), a 3-week residential programme (Day 0 to Day 21) and at-home follow up (D20 to D180). Clinical, biological and body composition parameters were measured at D0, D20, D90 and D180.

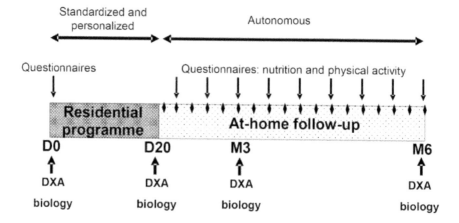

FIGURE 2: Study design: The two groups of volunteers (Normal and High protein-intake) followed a three-week residential programme with standardized and personalized diet and physical activity. Thereafter they returned home and were autonomous to manage their diet and physical activity, the latter being accompanied by a weekly session of physical activity on a voluntary basis. The two groups differed only by protein intake.

4.2.4 AT D0

Basal metabolic rate (BMR) was calculated by the equations of Black [27]: $BMR = 0.963 \cdot weight^{0.48} \cdot height^{0.50} \cdot age^{-0.13}$ for women, $BMR = 1.083 \cdot weight^{0.48} \cdot height^{0.50} \cdot age^{-0.13}$ for men. Anthropometrical and clinical values (weight, size height, BMI, waist circumference, blood pressure), body composition and biological parameters were measured. Before commencing the study, patients completed questionnaires concerning their food intake and PA over the previous week. Daily energy intake (DEI) and daily energy expenditure (DEE) were estimated from the self reported questionnaires. The food intake questionnaires identified possible deficiencies.

4.2.5 DURING THE THREE-WEEK RESIDENTIAL PROGRAMME

The subjects carried out daily individually adapted physical activities with a coach: walking (2 hours), aquagym (1 h, 3 times/week), keep fit activity (1 h, 3 times/week). Exercise intensity was fixed between 40 and 60% of the heart rate reserve (maximum theoretical heart rate – resting heart rate), by a heart rate recorder (Polar 4000). They followed the same programme between both groups six days a week. On the 7th day they only walked. Daily throughout the residential program, the patients received both standard and personalized balanced meals drawn up by dieticians. Their total daily food intake was calculated in order to reach a negative energy balance (EB = DEE – DEI) of 500 kcal/day. They also attended lectures on the MS, nutrition physiology, cooking and physical-activity. The aim of this educational support was to make them aware of the lifestyle they would need to adopt in order to maintain the beneficial effects of the regimen followed.

4.2.6 FROM D20 TO D180

The subjects returned home and were left in charge of managing the programme by themselves. They were asked to carry out the same training program and the same diet. Thereafter, PA sessions were organized once a week (keep fit activity or aquagym) to maintain compliance with the programme. They completed a questionnaire twice a month on their PA and eating habits. A dietician and a physical coach could be contacted if they had any queries.

4.2.7 METHODS

Clinical follow up was performed by a physician and psychological follow-up to verify the treatment acceptance was assessed by a psychologist. Daily energy intake (DEI) and physical activity index were based on questionnaires before and after the residential programme (three-day food intake and PA recorded once every 15 days). DEE was quantified by recording the time and intensity of each PA and the physical activity index (PAI=DEE/BMR) was calculated [28]. Each day during the residential programme, the patients received both standard and personalized balanced meals drawn up by dieticians. The aim was to restore macronutrient balance: 30 to 35% lipids, 15 to 20% proteins and carbohydrates for the rest. The PI difference of 0.2 g/kg/d between NPI and HPI represented, for a mean weight of 80 kg, about 16 g/d of protein per participant, a difference of 64 kcal/day. This difference was compensated for by the addition of the same quantity of carbohydrates for the NPI group. In both groups, daily PA was programmed for each subject to obtain a PAI equal to 1.4 [28,29].

4.2.8 BIOMETRY AND ANTHROPOMETRY

All subjects underwent medical examinations. Body height was measured with a stadiometer, BMI was calculated as the weight in kilograms divided by the square of the height in meters. Waist circumference and BP were recorded. Body composition was assessed by dual energy X-ray absorptiometry (DXA) (Hologic QDR Delphi series; Waltham, USA). The in vivo coefficients of variation were 4.2 and 0.48% for fat and lean mass, respectively. Central fat, (as a surrogate for visceral fat), was assessed by DXA, which measured the % fat in a rectangle from the upper edge of the second lumbar vertebra to the lower edge of the fourth lumbar vertebra. The vertical sides of this area were the continuation of the lateral sides of the rib cage [30]. All measurements for a given parameter were made by the same investigator.

4.2.9 BIOCHEMICAL MEASUREMENTS

Fasting blood samples were taken between 6.30 and 7.30 a.m., aliquoted and stored at −80°C until analyses. Basic biological examinations (glucose, lipid, creatinine, CRP and orosomucoid levels) were performed in the biochemistry laboratory of the University Hospital. Renal clearance was assessed by Cockcroft's formula [31].

4.2.10 STATISTICAL ANALYSIS

Gaussian distribution of the data was tested by the Kolmogorov-Smirnov test. Data are presented as means±standard error (SE). Significance was accepted at the $p < 0.05$ level. Statistical procedures were performed by SPSS Advanced Statistics software version 17 (SPSS Inc., Chicago, IL).

Using the method described by Howell [32], we calculated the number of subjects needed for a significant change in albumin levels between groups (based upon preliminary data). The minimum number was 10 participants per group for a $p < 0.05$ and a type II error of 10%. Under these conditions the statistical power was 90%.

Baseline characteristics were compared by analysis of variance or Fisher's exact test. Longitudinal changes between groups were tested with the use of mixed-model repeated-measures analysis of variance, with adjustment for baseline values and sex. The primary focus of the analyses was the 3-month change in albumin levels in the two groups.

In the event of interaction between the repeated measurements (time effect) and the main factor (group of protein intake), the changes within a group were analyzed either by a Newman-Keuls post-hoc test for normally distributed data, or a nonparametric Wilcoxon test for non-normally distributed data. Relationships between energy balance, physical activity and other data were assessed either by Pearson correlation or by multiple regression analysis.

4.3 RESULTS

4.3.1 DESCRIPTIVE CHARACTERISTIC OF PARTICIPANTS AT BASELINE

The descriptive characteristics of volunteers before the residential programme are presented in Table 1. There was no difference at baseline between groups. Usual food intake as indicated on questionnaires revealed that all patients of both groups had high lipid and low carbohydrate consumption, with a high ratio of high/low glycemic index (Table 1). Mean cholesterol consumption was 341 ± 97 mg/d. The mean DEE before the residential programme was 1983 ± 228 kcal/day corresponding to a low PAI of 1.22 ± 0.09.

TABLE 1: Baseline characteristics of participants*

Characteristic	Normal protein intake (n=14)	High protein intake (n=14)	Significance
Age – years	62.9 ± 6.9	60.6 ± 6.0	NS
Male (M) : n (%)	10 (71)	9 (64)	NS
Female (F) : n (%)	4 (29)	5 (36)	NS
Weight (kg) – M	90.4 ± 8.7	96.3 ± 4.1	NS
Weight (kg) - F	90.0 ± 18.7	88.7 ± 14.5	NS
BMI (kg/m²)	32.1 ± 4.2	35.2 ± 4.2	NS
MS parameters			
BMI (kg/m²)	32.1 ± 4.2	35.2 ± 4.2	NS
Waist circumference (cm) – M	106.5 ± 7.7	104.3 ± 18.5	NS
Waist circumference (cm) – F	101.5 ± 5.8	100.6 ± 9.2	NS
Blood Pressure (mmHg)	$137/83 \pm 11/5$	$135/86 \pm 18/13$	NS
Triglycerides (mmol/l)	1.68 ± 1.15	1.88 ± 0.55	NS
HDL (mmol/l)	1.44 ± 0.42	1.06 ± 0.33	NS
Glycemia (mmol/l)	6.15 ± 1.86	5.13 ± 0.68	NS
Use of medication: n (%)			

TABLE 1: *Cont.*

Characteristic	Normal protein intake (n=14)	High protein intake (n=14)	Significance
Antihypertensive agents	5 (36)	6 (29)	NS
Lipid lowering drugs	3 (21)	4 (29)	NS
Hypoglycemiant drugs	3 (21)	5(36)	NS
Basal Metabolic Rate	1626±178	1587±254	NS
Daily Energy Expenditure (kcal/d)	1983±229	1920±307	NS
Physical Activity Index	1.22±0.09	1.19±0.12	NS
Daily Energy Intake (kcal/d)	2073±556	1921±348	NS
Daily Energy Intake (kcal/kg/d)	23.8±6.5	21.2±4.9	NS
Percentage of each macronutrient in the food			
% Carbohydrates	39.2±5.5	40.5±6.0	NS
of high glycemic index carbohydrates	13.9±5.3	13.0±1.2	NS
% Lipids	44.5±4.7	41.9±4.4	NS
% Proteins	16.3±1.5	17.6±2.7	NS
Protein intake (g/kg/d)	0.91±0.26	0.90±0.22	NS

*: *Plus–minus values are means ±SD.*

4.3.2 DESCRIPTIVE CHARACTERISTICS DURING THE RESIDENTIAL PROGRAMME

4.3.2.1 INTERVENTION

Food intake over the six months is presented in Table 2, and energy balance and PAI in Table 3. DEI decreased and DEE increased during the residential programme, both significantly, resulting in a negative balance. The PAI was set at 1.4±0.1. Macronutrient distribution improved: lipid intake was lower and that of carbohydrates and proteins higher. There were reduced amounts of saturated fatty acids and cholesterol (Table 2).

TABLE 2: Changes in food intakes: percentage of each macronutrient (carbohydrates, lipids, proteins)

Variable	Groups	D0	D20	D90	D180
% Carbohydrates	NPI	39.2±5.5	48.3±2.5†	42.5±4.6†	39.1±3.4
	HPI	40.5±6.0	47.3±3.2†	41,6±4.6	39.5±4.0
% Lipids	NPI	44.5±4.7	32.8±2.1†	38.2±4.7	41.7±3.6
	HPI	41.9±4.4	28.7±3.4†	35.5±5.2	38.1±5.6
% Proteins	NPI	16.3±1.5	18.9±0.8*	19.4±0.19†*	19.4±1.3†*
	HPI	17.6±2.7	24.8±1.6†*	23.0±2.3†*	22.2±2.2†*
Protein Intake (g/kg/d)	NPI	0.91±0.26	0.95±0.11*	0.94±0.19*	0.96±0.17*
	HPI	0.90±0.22	1.19±0.13†*	1.10±0.14†‡*	1.09±0.20†‡*

*All calculations were done on Bilnut program using CIQUAL (S.C.D.A. Nutrisoft, Le Hallier 37390 Cerelles, France). * : $p < 0.05$ to compare the percentage change between D0 and D20, D90 and D180 in the two groups (High Protein Intake versus Normal Protein Intake) using the Bonferroni test. † : $p < 0.05$ to compare the value at the follow-up time with the baseline value (D0) within each group, as calculated by mixed-model repeated-measures analysis of variance. ‡ : $p < 0.05$ to compare the value at the follow-up time with the end of the residential programme (D20) within each group, as calculated by mixed-model repeated-measures analysis of variance.*

4.3.2.2 GENERAL EFFECTS FOR BOTH GROUPS

At the end of the residential programme, the combination of diet and PA had significantly reduced body weight, BMI, waist circumference and systolic BP. Fifty seven percent of weight loss was in fat mass, with a significant decrease in central fat. The rates of HDL remained stable and triglyceride levels had an overall tendency to decrease ($p < 0.1$). Other blood lipid parameters decreased significantly at D20 and then returned to baseline levels. Creatinine levels and creatinine clearance, as assessed by Cockroft's formula remained stable (Table 3).

Following the residential programme, dietary recommendations were given every 15 days to maintain a negative energy balance. The patients gradually went back to their former eating habits with increased lipid intake and a reduction in carbohydrates. Weight loss and central fat loss continued, as did the reduction in BMI and waist circumference, with no significant difference between groups (Table 3).

TABLE 3: Changes in energy balance, physical activity index, MS parameters and body composition during the residential programme and follow-up for the two groups of protein intake (PI): Normal (NPI) and High (HPI) with respectively a PI at 1.0 g/kg/day and 1.2 g/kg/day

Variable	Groups	D0	D20	D90	D180
Body composition measured by DEXA:					
Weight (kg)	NPI	90.3±11.6	86.5±11.0†	84.4±11.1†‡	85.4±12.1†‡
	HPI	94.1±15.2	90.7±14.5†	87.1±13.6†‡	86.4±15.0†‡
BMI (kg/m²)	NPI	32.1±4.2	30.7±3.8†	29.9±3.5†‡	30.3±3.7†‡
	HPI	35.2±4.2	33.9±4.1†	32.6±4.3†‡	32.4±4.7†‡
Lean (kg)	NPI	58.287±7.47	57.323±7.56†	56.488±7.89†	56.798±7.99†
	HPI	56.594±11.04	55.987±10.32	55.223±10.57†	55.137±10.47†
Total fat (kg)	NPI	29.542±9.75	26.737±9.38†	25.524±8.94†	26.213±9.580†
	HPI	35.229±8.25	32.474±7.72†	29.598±7.68†‡	29.068±9.18†‡
Visceral fat (kg)	NPI	3.277±1.24	2.839±1.22†	2.695±1.07†	2.527±1.01†‡
	HPI	3.295±0.88	2.916±0.79†	2.662±0.75†	2.445±0.71†‡
Fat percentage (%)	NPI	32.3±7.9	30.5±8.3†	29.9±8.2†	30.2±8.3†
	HPI	37.4±6.1	35.7±5.9†	34.0±6.6†	33.4±7.2†
Energy Balance (kcal/d)	NPI	+92±521	−751±147†	−521±304†‡	−413±304‡
	HPI	+34±348	−635±102†	−524±83†‡	−444±71‡
Physical Activity	NPI	1.22±0.09	1.42±0.07†	1.31±1.97†‡	1.30±0.07†‡
	HPI	1.19±0.12	1.40±0.09†	1.33±1.82†‡	1.28±0.04†‡
Metabolic Syndrome parameters:					
Waist circumference (cm)	NPI	105.1±7.4	102.7±7.9†‡	99.2±6.3†‡	98.9±7.4†‡
	HPI	101.6±11.8	97.2±9.6†‡	95.1±9.9†‡	93.3±9.1†‡
Blood Pressure (mmHg)	NPI	137/83±11/5	130†/80±13/5	129†/80±15/6	132/78±16/9
	HPI	135/86±19/13	128†/80±15/16	125†/85±16/18	127/82±20/16
Triglycerides (mmol/l)	NPI	1.68±1.15	1.19±0.34	1.17±0.40	1.27±0.61
	HPI	1.88±0.55	1.37±0.26	1.85±0.21	1.86±1.07
HDL (mmol/l)	NPI	1.44±0.42	1.49±0.31	1.31±0.31‡	1.58±0.46
	HPI	1.06±0.33	1.02±0.29	1.13±0.33	1.14±0.25
Glycemia (mmol/l)	NPI	6.15±1.86	6.26±1.76	5.7±1.97‡	6.28±2.34
	HPI	5.13±0.68	4.54±0.54†	4.98±0.79	4.92±0.50

TABLE 3: *Cont.*

Variable	Groups	D0	D20	D90	D180
Other Lipid parameters:					
Total cholesterol (mmol/l)	NPI	6.08±1.46	5.07±0.95†	5.13±1.25†	5.56±0.89
	HPI	5.79±1.07	4.65±1.09†	6.18±1.60	6.19±1.08
LDL (mmol/l)	NPI	3.79±1.21	3.05±0.87†	3.47±1.25	3.43±0.95
	HPI	3.89±1.01	2.99±0.99†	4.21±1.34	4.25±0.97
Albumin levels (g/l)	NPI	40.6±3.3	40.3±3.1	34.3±1.8†‡*	35.3±2.2†‡
	HPI	40.8±2.4	41.3±2.6	41.5±2.6*	41.0±3.2*
Pro-inflammatory factors:					
CRP (mg/l)	NPI	5.09±4.06	3.64±3.69	4.18±5.60	3.68±4.12
	HPI	4.19±2.33	3.80±3.48	3.04±3.23	2.99±2.05
orosomucoid (mg/l)	NPI	0.85±0.16	0.80±0.19	0.75±0.19	0.84±0.15
	HPI	0.91±0.29	0.80±0.33	0.91±0.25	0.84±0.41
Renal function:					
creatinine levels (mmol/l)	NPI	89.4±19.9	89.1±18.8	80.1±20.7	81.1±20.4
	HPI	84.9±22.2	88.2±24.1	86.1±23.7	84.0±21.3
Cockroft (ml/min)	NPI	100.3±28.3	95.4±24.2	107.1±27.9	106.6±32.0
	HPI	111.8±23.7	104.5±24.9	106.7±23.4	109.0±24.4

* : $p < 0.05$ to compare the percentage change between D0 and D20, D90 and D180 in the two groups (High Protein Intake versus Normal Protein Intake) using the Bonferroni test. † : $p < 0.05$ to compare the value at the follow-up time with the baseline value (D0) within each group, as calculated by mixed-model repeated-measures analysis of variance. ‡ : $p < 0.05$ to compare the value at the follow-up time with the end of the residential programme (D20) within each group, as calculated by mixed-model repeated-measures analysis of variance.

4.3.3 MAIN JUDGMENT CRITERIA

At D20, there was no significant change in albumin levels in either group. Following the residential programme, the levels decreased in NPI group at D90 and D180, to reach the threshold value of 35 g/l considered as a marker of a protein deficiency. Albumin levels in the HPI group remained stable throughout the experimental period.

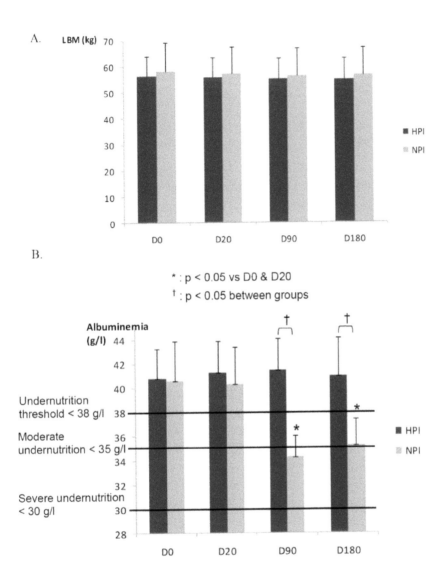

FIGURE 3: Change from baseline (day 0) to Month 6 (day 180) in albumin levels and lean body mass (LBM) for both groups: normal protein intake (NPI) set at 1.0 g/kg/d and high protein intake (HPI) at 1.2 g/kg/d. Solid bars represent albumin levels or lean mass in High protein intake group. Open bars represent albumin levels or lean mass in Normal protein intake group. T bars indicate standard errors. Panels A and B show the change in lean mass and albumin levels, respectively, for all participants (n= 28), who were randomly assigned to a High protein intake (n= 14) or to a Normal protein intake (n= 14). No missing data.

4.3.4 SECONDARY JUDGMENT CRITERIA

In both groups, creatinine levels and creatinine clearance remained stable at all times, as did CRP and orosomucoid.

4.4 DISCUSSION

Our study shows that when PA and nutritional habits are modified by a healthier lifestyle, there is a significant improvement in MS criteria (Table 3). When subjects entered the study, protein intake was about 0.9 g/kg/d, the level currently considered as adequate for the senior [10] and albumin levels were normal (Table 3, Figure 3). As the exercise regimen and the reduced caloric intake of the programme were expected to increase the protein LTI, we set protein intake at 1 g/kg/d for NPI and at 1.2 g/kg/d for HPI.

We measured albumin levels at the beginning of the study (D0) and at D20, D90 and D180 in the two groups of MS patients older than 50 years. There was no change in the levels in the HPI group, while in the NPI group, they were lower at D90. This significant decrease would probably indicate a too low protein intake in the NPI group, given the new conditions of PA and overall caloric intake [9,12-14]. In contrast, the PI set at 1.2 g/kg/d kept albumin levels steady. The fall in level was observed at D90 only and not at D20, probably due to its long half-life of 20 days.

A catabolic phase may occur in the event of increased inflammatory status [33]. In the present study, we monitored this status by assaying CRP, which remained stable with normal levels in both groups (Table 3). Likewise, altered renal function may be the cause of hypoalbuminemia. This was not the case in our patients, as assessed by normal and stable renal Cockcroft clearance (Table 3). Moreover, high protein intake is now considered to be only a weak risk for renal function in healthy senior individuals [34]. Finally, all patients were free of medications which may influence albumin levels.

Consequently, our study shows that LTI (and not RDA) for protein must be set at 1.2 g/kg/d when physical activity together with reduced caloric intake are prescribed to patients suffering from MS. This result agrees with the findings of Lucas & Heiss [15], who proposed a RDA of

1–1.3 g/kg/d for older adults (> 50 years old) engaged in physical train-ing corresponding to a LTI of 0.8 to 1.0 g/kg/day. The addition of caloric restriction, as in this study, increases LTI for protein to 1.2 g/kg/d.

Such a level of protein intake may appear high for people consuming light meals. For an individual weighing 80 kg, it represents 100 g/day of protein dry weight, that is to say 500 g of crude protein. Dietary animal protein is the primary source of high biological value protein [35]. If fifty percent of protein intake is from animal origin, this corresponds to 250 g meat or fish per day, since eggs and cheese are drastically reduced or even suppressed on account of their high lipid content. We may also consider that 100 g protein represent 400 kcal. If the total caloric intake is set at 2000 kcal, proteins will represent 20% of intake.

4.5 CONCLUSION

This study is a contribution to the quantification of the optimal protein lower threshold intake for individuals suffering from metabolic syndrome entering a programme of weight reduction with controlled diet and exer-cise. This is an important issue because insufficient protein intake could be detrimental when physical activity is added to a restricted diet. Our study suggests protein intake should be 1.2 g/kg/d in senior people suffer-ing from metabolic syndrome and taking part in a weight loss programme.

REFERENCES

1. Alberti KG, Zimmet P, Shaw J: IDF epidemiology task force consensus group: the metabolic syndrome: a new worldwide definition. Lancet 2005, 366:1059-1062.
2. Bassand JP: Managing cardiovascular risk in patients with metabolic syndrome. Clin Cornerstone 2006, 8:S7-S14.
3. Laaksonen DE, Lakka HM, Niskanen LK, Kaplan GA, Salonen JT, Lakka TA: Meta-bolic syndrome and development of diabetes mellitus: application and validation of recently suggested definitions of the metabolic syndrome in a prospective cohort study. Am J Epidemiol 2002, 156:1070-1077.
4. Ford ES, Giles WH, Dietz WH: Prevalence of the metabolic syndrome among US adults: findings from the third National Health and Nutrition Examination Survey. JAMA 2002, 287:356-359.

5. Rosenberg IH: Sarcopenia. Origins and clinical relevance. J Nutr 1997, 127:990S-991S.
6. Hughes VA, Frontera WR, Wood M, Evans WJ, Dallal GE, Roubenoff R, Fiatarone Singh MA: Longitudinal muscle strength changes in older adults: influence of muscle mass, physical activity, and health. J Gerontol A Biol Sci Med Sci 2001, 56:B209-B217.
7. Guillet C, Prod'homme M, Balage M, Gachon P, Giraudet C, Morin L, Grizard J, Boirie Y: Impaired anabolic response of muscle protein synthesis is associated with S6K1 dysregulation in elderly humans. FASEB J 2004, 18:1586-1587.
8. Hébuterne X, Broussard JF, Rampal P: Acute renutrition by cyclic enteral nutrition in elderly and younger patients. JAMA 1995, 273:638-643.
9. Martin A: Apports Nutritionnels Conseillés pour la population française. 3e edition. Paris: Tec & Doc Lavoisier; 2000.
10. Trumbo P, Schlicker S, Yates AA, Poos M: Dietary reference intakes for energy, carbohydrate, fiber, fat, fatty acids, cholesterol, protein and amino acids. J Am Diet Assoc 2002, 102:162-30.
11. Paddon-Jones D, Short KR, Campbell WW, Volpi E, Wolfe RR: Role of dietary protein in the sarcopenia of aging. Am J Clin Nutr 2008, 87:1562S-1566S.
12. American College of Sports Medicine; American Dietetic Association; Dietitians of Canada: Joint position statement: nutrition and athletic performance. Med Sci Sports Exerc 2000, 32:2130-2145.
13. Zello GA: Dietary reference intakes for the macronutrients and energy: considerations for physical activity. Appl Physiol Nutr Metab 2006, 31:74-79.
14. Evans WJ: Protein nutrition, exercise and aging. J Am Coll Nutr 2004, 23:601S-609S.
15. Lucas M, Heiss CJ: Protein needs of older adults engaged in resistance training: a review. J Aging Phys Act. 2005, 13:223-236.
16. Fuller MF, Crofts RM: The protein-sparing effect of carbohydrate. 1. Nitrogen retention of growing pigs in relation to diet. Physiol Rev 1951, 31:449-488.
17. Munro HN: Carbohydrate and fat as factors in protein utilization and metabolism. Br J Nutr 1977, 38:479-488.
18. Pellet PL, Young VR: The effects of different levels of energy intake on protein metabolism and of different levels of protein intake on energy metabolism: a statistical evaluation from the published literature. In Protein energy interactions. IDECG, Waterville valley, NH Edited by Scrimshaw NS, Schürch B. 1991, 81.
19. Fry CS, Rasmussen BB: Skeletal muscle protein balance and metabolism in the elderly. Curr Aging Sci 2011, 4:260-268.
20. Fouque D, Guebre-Egziabher F: Do low-protein diets work in chronic kidney disease patients? Semin Nephrol 2009, 29:30-8.
21. Larosa JC, Fry AG, Muesing R, Rosing DR: Effects of high-protein, low-carbohydrate dieting on plasma lipoproteins and body weight. J Am Diet Assoc 1980, 77:264-70.
22. Lindeman RD, Tobin J, Shock NW: Longitudinal studies on the rate of decline in renal function with age. J Am Geriatr Soc 1984, 33:278-285.
23. Lindeman RD, Tobin JD, Shock NW: Association between blood pressure and the rate of decline in renal function with age. Kidney Int 1984, 26:861-8.

24. Kasiske BL: Relationship between vascular disease and ageassociated changes in the human kidney. Kidney Int 1987, 31:1153-9.

25. Melchior JC: How to assess preoperative nutritional status? Ann Fr Anesth Reanim 1995, 14:19-26.

26. Koretz RL: Death, morbidity and economics are the only end points for trials. Proc Nutr Soc 2005, 64:277-84.

27. Black AE, Coward WA, Cole TJ, Prentice AM: Human energy expenditure in affluent societies: an analysis of 574 doubly-labelled water measurements. Eur J Clin Nutr 1996, 50:72-92.

28. Ainsworth BE, Haskell WL, Leon AS, Jacobs DR, Montoye HJ, Sallis JF, Paffenbarger RS: Compendium of physical activities: classification of energy costs of human physical activities. Med Sci Sports Exerc 1993, 25:71-80.

29. Pannemans DL, Westerterp KR: Energy expenditure, physical activity and basal metabolic rate of elderly subjects. Br J Nutr 1995, 73:571-81.

30. Kamel EG, McNeill G, Van Wijk MC: Usefulness of anthropometry and DXA in predicting intra-abdominal fat in obese men and women. Obes Res 2000, 8:36-42.

31. Cockcroft DW, Gault MH: Prediction of creatinine clearance from serum creatinine. Nephron 1976, 16:31-41.

32. Howell DC: Statistical methods. Paris: De Boek Université; 1998.

33. Bonnefoy M, Laville M, Ecochard R, Jusot JF, Normand S, Maillot S, Lebreton B, Jauffret M: Effects of branched amino acids supplementation in malnourished elderly with catabolic status. J Nutr Health Aging 2010, 14:579-84.

34. Millward DJ: Optimal intakes of protein in the human diet. Proc Nutr Soc 1999, 58:403-13.

35. Chernoff R: Protein and older adults. J Am Coll Nutr 2004, 23:627S-30.

CHAPTER 5

A Prospective Study of Nutrition Education and Oral Nutritional Supplementation in Patients with Alzheimer's Disease

GLAUCIA A. K. PIVI, ROSIMEIRE V. DA SILVA, YARA JULIANO, NEIL F. NOVO, IVAN H. OKAMOTO, CЙSAR Q. BRANT, AND PAULO H. F. BERTOLUCCI

5.1 BACKGROUND

Weight loss in patients with Alzheimer's disease (AD) is a common clinical manifestation. In these patients, impaired nutritional status results in changes in body composition and biochemistry indicators [1]. In order to identify such weight loss, some studies correlated organic deficiency and low energy intake or hypercatabolism in patients, thus suggesting that weight loss may be a risk factor in the etiology of dementias and other psychiatric and cognitive disorders, however this has not been evaluated [2-4]. Higher infection rates, increased energy expenditure due to repetitive movements, and cognitive deficit impairing AD patient independence may also be considered as a cause of weight loss [5,6]. Weight loss increases

A Prospective Study of Nutrition Education and Oral Nutritional Supplementation in Patients with Alzheimer's Disease . © Pivi GAK, da Silva RV, Juliano Y, Novo NF, Okamoto IH, Brant CQ, and Bertolucci PHF. Nutrition Journal *10,98 (2011). doi:10.1186/1475-2891-10-98. Licensed under a Creative Commons Attribution 2.0 Generic License, http://creativecommons.org/licenses/by/2.0.*

the risk for infections, pressure ulcer development and poor wound heal-
ing, which in turn can impair AD patient's quality of life [7].

Some strategies can be adopted to improve the nutritional status of
these patients. These strategies include patient nutrition education programs, and the use of oral nutritional supplements, which can significantly
impact nutritional status [8-10]. Thus, the need to study strategies of nutritional intervention which minimize or improve AD patients' nutritional
status is justified, supporting the conduction of this study. The objective
of this study is to determine if there is any difference between oral nutritional supplementation and nutrition education on the nutritional status of
patients with AD.

5.2 METHODOLOGY

A randomized, 6-month, prospective study was conducted at the clinic
of the Behavioral Neurology Sector, in the Neurology and Neurosurgery
Department of Universidade Federal de São Paulo - Escola Paulista de
Medicina (UNIFESP/EPM).

The sample consisted of 90 subjects, of both genders, aged at least
65 years old and with probable AD, according to Diagnostic Statistical
Manual 4th Edition (DSM IV) criteria and clinical dementia rating (CDR)
of 1, 2 or 3.

For exclusion criteria were considered other forms of dementia, alternative feeding requirement (tube feeding), type 1 and 2 diabetes mellitus,
and renal diseases.

Twelve subjects were included in the study but not in the statistical
analys is: 3 subjects from CG and 4 from EG had difficulty in being transported to the hospital; 3 subjects from SG and 1 from CG died; 1 subject
from SG needed tube feeding.

The remaining subjects were randomized into 3 groups: control group
(CG), (N = 27), education group (EG), (N = 25), and supplementation
group (SG), (N = 26).

All subjects were assessed at baseline and then at monthly intervals
during the 6-month study period, including an orientation of health nutrition. The subjects nutritional status was assessed using anthropomet-

ric and biochemical data. The anthropometric data included: height (m), current weight (kg), Body Mass Index (BMI) (kg/m^2), arm circumference (AC) (cm), arm muscle circumference (AMC) (cm), and triceps skinfold (TSF) (mm). For weight were used Welmy® mechanical scales for adults with a 150 kg capacity and for height was utilized stadiometer graduated in centimeters, to for subjects that were able to maintain erect posture. For the others with presented posture problems as kyphosis or lordosis and were usable to stand, the stature and knee equation proposed by Chumlea (1985) [11] was used, to avoid biases in measures of statures. For the compartmental muscle and fat mass assessment, Lange® Skinfold calipers were used, which expresses results in millimeters (mm). The biochemistry data included: total protein (TP), serum albumin and total lymphocyte count (TLC). Biochemical data were collected following a 12-hour fast and evaluated by the central laboratory of Hospital São Paulo.

For cognitive tracking, all subjects were assessed using the Mini-Mental State Examination (MMSE) assessing temporal and spatial orientation, memory, attention, calculation, language and praxis [12,13] and Clinical Dementia Rating (CDR) [14,15] administered by a duly qualified neuropsychologist.

The following demographic data were also collected for all subjects: education level (years of schooling), time of disease evolution calculated from the date of diagnosis and dependence during meals.

Subjects in the CG were monitored by monthly nutritional assessments and did not receive any form of intervention. The 25 Subjects caregivers and patients, in the EG participated in the educational program which consisted of 10 classes. Each class was taught to a maximum of 10 participants, with the aim greater interaction between the professional and caregivers. Each expositive class was supported by slides, with themes were of data proposed in accordance with the Brazilian Association of Alzheimer's (ABRAZ) Caovilla & Canineu (2002) [16] a nonprofit association that assists caregivers and family members of patients with Alzheimer's disease. Also it took into account the information received from the caregivers based on the main nutrition deficits in AD patients. The classes were developed with relevant topics to the needs of nutritional intervention, such as: the importance of nutrition in disease, behavioral changes during meals, attractive meals, constipation, hydration, administration of drugs, swallowing, food supplementation, lack of appetite, clarification of doubts. This order of exposition follows the progression of AD and the onset of symptoms that may be related to nutrition. Subjects in the SG received oral nutritional supplementation twice daily for 6 months in addition to their usual diet (Ensure with FOS®, Abbott Nutrition) and were measured monthly by means of anthropometry and biochemical parameters, as described before. Two servings provide 680 kcal and 25.6 g protein/day.

All subjects or their representatives signed the Informed Consent Form. This research was approved by the Ethics Committee of Universidade Federal de São Paulo (protocol 0552/06).

Statistical analysis was undertook using Kruskal-Wallis variance analysis to compare the three groups in relation to anthropometric, biochemistry, demographic and disease stage variables, and Siegel's chi-square test to study possible associations between the groups and the variables studied. The rejection level of the null hypothesis was fixed at $\alpha = 0.05$.

5.3 RESULTS

Ninety (90) subjects were enrolled in the study, and 86.67% (n = 78) completed the study. Of the subjects, 30% (n = 27) were in the CG; 27.78% (n = 25) were in the EG; and 28.88% (n = 26) were in the SG. 8.89% (n = 8) of the patients did not meet the inclusion criteria and 4.44% (n = 4) died.

Of the 78 subjects who completed the study, 67.9% (n = 53) were female and 32.1% (n = 25) were male, with mean age of 75.2 years and median age of 76 years.

As shown in tables 1 and 2, 39.7% (n = 31) were in the moderate phase of the disease (CDR 2) and had a mean MMSE of 12.32, however there were no statistically significant differences among the groups. Demographic data (table 1) as well as dependence during meals (table 2) were also not statistically different among the groups. All these variables show that the groups were similar and did not have any influence on the results obtained.

TABLE 1: Demographic and MMSE data of Alzheimer's Disease (AD) patients, according to Control Group (CG), Education Group (EG) and Supplementation Group (SG)

	Control		Education		Supplementa-tion		Kruskal-Wallis variance analysis		
	Mean	Median	Mean	Median	Mean	Median	Calcu-lated H	P	Signifi-cance
MMSE	12.59	18.00	12.80	12.00	11.58	9.00	0.14	0.932	N.S.
Age	75.22	74.00	75.88	76.00	76.38	78.00	0.61	0.738	N.S.
School.	3.44	3.00	5.40	4.00	4.61	4.00	3.62	0.164	N.S.
Tempev.	5.93	5.00	6.40	6.00	6.19	6.00	0.45	0.798	N.S.

MMSE: mini-mental state examination; School.: schooling in years; TEMPEV.: time of disease evolution; p: significance level; N.S.: not significant.

TABLE 2: Patients with Alzheimer's Disease (AD) in Control Group (CG), Education Group (EG) and Supplementation Group (SG) compared for CDR, gender and dependence during meals

	CDR			Gender		Dependence during Meals	
	1 (n)	2 (n)	3 (n)	Male (n)	Female (n)	Not dep.	Dep.
CG	9	9	9	7	20	14	13
EG	7	11	7	8	17	17	8
SG	7	11	8	10	16	13	13
Total	23	31	24	25	53	44	34

Critical $X^2 = 5.99$ calculated X^2 for CDR $= 0.77$ not significant. Critical $X^2 = 5.99$ calculated X^2 for gender $= 0.96$ not significant. Critical $X^2 = 5.99$ calculated X^2 for dependence during meals $= 2.03$ not significant

Regarding the anthropometric and biochemical parameters (Table 3), the SG showed a significant improvement in the anthropometric data of weight (H calc $= 22.12$, p $=< 0.001$), body mass index (H calc $= 22.12$, p $=< 0.001$), arm circumference (H calc $= 12.99$, p $= 0.002$), and arm muscle circumference (H calc $= 8.67$, p $= 0.013$) compared to CG and EG after 6 months. There were no statistically significant differences in TSF among the three groups after 6 months. The BMI in the EG was significant compared to CG. There were significant differences in total protein (H calc = 6.17, p = 0.046) in the SG compared to the other groups. The total lymphocyte count in the SG and EG was significant compared to the CG (H cal = 7.94, p = 0.019) after 6 months.

There were no statistically significant differences in serum albumin among the three groups.

5.4 DISCUSSION

Most subjects studied had CDR 2, which is relevant since most feeding behavior changes, such as forgetfulness or feeding voracity, are seen in this phase of the disease [17].

In the EG, BMI showed a significant increase compared to CG (p < 0.001). Likewise, biochemical (TP and TLC, p = 0.046, p = 0.019 respectively) and anthropometric indicators (AC and AMC, p = 0.002 and p = 0.013 respectively) also improved in EG group. Our results support the findings from other studies, showing that nutritional education has a positive effect on the diet and modifies nutrition knowledge [18,7,19].

There were significant differences in the SG's anthropometric measures (current weight, BMI, AC, AMC) compared to CG and EG, showing that the use of oral nutritional supplementation improves the patient's nutritional status. In addition, since there were no significant differences in TSF measurement among the three groups, and particularly in the SG, showed that there was no increase in subcutaneous body fat. Furthermore, the oral nutritional supplementation did not result in significant difference in TSF measurement among the three groups, indicating that additional oral nutritional supplementation did not increase subcutaneous body fat in this sample.

Oral nutritional supplementation provided to dementia patients significantly improves nutritional status and the quality of the diet consumed [20,21], but despite these results, Trelis & López [22] observed that only 11% of outpatients used oral nutritional supplements.

TABLE 3: Mean and median values of anthropometric and biochemical measures for patients with Alzheimer's disease in the different groups

	Control		Education		Supplementation		Kruskal-Wallis		
	Mean	Median	Mean	Median	Mean	Median	Calculated H	p	Significance
Height	1.57	1.57	1.54	1.54	1.58	1.56	02.20	0.333	N.S.
CW	-2.20 (61.87 - 60.65)	-1.81 (57.60 - 55.70)	1.19 (54.29 - 50.00)	0.32 (54.54 - 51.70)	6.66 (54.70 - 57.93)	4.37 (53.55 - 56.80)	22.12	<0.001	S > C and E
BMI	-2.21 (24.81 - 24.32)	-1.82 (23.55 - 22.80)	1.19 (22.71 - 22.84)	0.31 (23.82 - 23.90)	6.55 (21.66 - 22.98)	4.37 (22.38 - 23.15)	21.94	<0.001	S > C and E
AC	-0.41 (26.14 - 26.07)	0.00 (26.40 - 26.30)	1.87 (24.72 - 25.11)	2.34 (25.60 - 26.00)	5.44 (23.99 - 25.20)	4.76 (23.90 - 24.80)	12.99	0.002	S > C and E
TSF	2.20 (15.67 - 15.85)	0.00 (15.00 - 16.00)	2.32 (14.20 - 16.40)	1.00 (14.00 - 16.00)	1.44 (11.54 - 13.55)	1.58 (12.00 - 14.00)	3.98	0.136	N.S.
AMC	-0.19 (21.21 - 21.60)	-0.90 (21.12 - 21.01)	-1.27 (20.25 - 19.96)	-2.06 (19.80 - 19.66)	3.43 (20.36 - 21.01)	2.54 (20.44 - 20.93)	8.67	0.013	S > C and E
Total prot.	0.09 (06.94 - 06.95)	0.00 (07.00 - 06.90)	-1.04 (06.55 - 06.84)	-1.39 (06.80 - 06.80)	4.30 (06.46 - 07.02)	3.28 (06.8 - 07.00)	6.17	0.046	S > C and E
Albumin	-3.15 (04.31 - 04.17)	-4.35 (04.30 - 04.10)	-4.06 (04.05 - 04.08)	-4.54 (04.20 - 04.20)	0.69 (03.81 - 04.07)	0.00 (04.20 - 04.20)	2.54	0.281	N.S.

TABLE 3: *Cont.*

	Control		Education		Supplementation		Kruskal-Wallis		
	Mean	Median	Mean	Median	Mean	Median	Calculated H	p	Significance
TLC	13.43 (2045.24 - 2279.46)	10.50 (2034.35 - 2416.04)	-1.06 (1888.85 - 1835.25)	-1.41 (1852.40 - 1806.12)	12.57 (1464.20 - 1808.68)	10.68 (1541.10 - 1582.50)	7.94	0.019	S and E > C

CW = Current Weight (kg); BMI = Body Mass Index (Kg/m2); AC = Arm Circumference (cm); TSF = Triceps Skin fold (mm) and AMC = Arm Muscle Circumference (cm); Total prot. = total serum protein (µg/dl); TLC = Total Lymphocyte Count (mm3); p: significance level; N.S. = not significant

In Brazil, there are no studies of outpatients using any type of oral nutritional supplements. However, this study clearly indicates that it is important for all health care professionals involved in the treatment of patients with AD to be able to detect the presence of nutritional deficits. This enables health care professionals to refer the patient to a qualified health professional, so the best nutritional intervention can be implemented.

The benefits of using oral nutritional supplementation have also been studied in elderly subjects without dementia and have shown their efficacy in addition to usual diet. The addition of 500 kcal/day via oral nutritional supplementation for elderly has been shown to improve convalescence and recovery from deficiency states [23,24].

In relation to the biochemical parameters used to assess nutritional status, significant differences were only seen for total protein in the SG and total lymphocyte count in the SG and EG.

Increased total protein in the SG was not correlated with the use of oral nutritional supplementation, but this result may be correlated with evolution of the disease itself. This parameter could perhaps be better assessed with a larger sample or longer follow up.

Increased total lymphocyte count (TLC) in the EG suggests that nutrition education might contribute to the improvement in the immune status of the patients. This may be due to the influence of education on the choice of healthier food, which contributes to the nutritional status as a whole [25].

On the other hand, the significant improvement in TLC in SG subjects showed that the use of oral nutritional supplementation improves immune status. This may be due to the antioxidant micronutrients in the supplement, particularly selenium, β-carotene, vitamin C and E, which impact with the immune system [26,27].

Improving or maintaining the nutritional status of patients with AD should be a priority of patient treatment. Therefore this is a responsibility for the entire multiprofessional team.

Implementing nutrition education improved the BMI of patients with AD, and thus its use is viable for any kind of healthcare service due to its low cost and positive impact on the modification of dietary habits. In addition, oral nutritional supplementation should be part of the usual diet of these patients, as the additional nutrients provided contribute to an improved nutritional status [28,29].

5.5 CONCLUSION

Oral nutritional supplementation was shown to be more effective compared to nutrition education in improving nutritional status of patients with Alzheimer's Disease.

REFERENCES

1. Contreras AT: Nutrición em la enfermedad de Alzheimer. Arch Neurocien 2004, 3(9):151-158.
2. Barrett-Connor E, Edelstein SL, Corey-Bloom J, Jernigan T, Archibald S, Thal LJ: Wheight loss precedes dementia community-dwelling older adults. J Am Geriatr Soc 1996, 44:1147-1152.
3. Reyes-Ortega G, Guyonnet S, Ousset PJ, Nourhashemi F, Vellas B, Albarède JL, De Glizezinski I, Riviere D, Fitten LJ: Weight loss in Alzheimer's disease and resting energy expenditure (REE), a preliminary report. J Am Geriatr Soc 1997, 45(11):1365-70.
4. Poehlman ET, Dvorak RV: Energy expenditure, energy intake, and weight loss in Alzheimer disease. Am J Clin Nutr 2000, (71):650S-655S.
5. DiLuca DuW, Growden JH: Weight loss in. Alzheimer's Disease. J Geriatric Psychiatry Neurol 1993, 6:34-38.
6. Kikuchi EL, Filho WJ: Tratamento e prevenção de Distúrbios Físicos associados. In Demência Abordagem Multidisciplinar. Edited by Caixeta L. São Paulo: Atheneu; 2006:531-538.
7. Riviere S: Program of nutritional education in the prevention of weight loss and slow cognitive decline in the disease of Alzheimer. J Nutr Health Aging 2001, (5):295-99.
8. Turano W, Almeida CCC: Educação Nutricional. In: Gouveia ELC. In Nutrição Saúde e Comunidade. 2nd edition. Rio de Janeiro: Revinter; 1999:57-60.
9. Burns BL, Davis EMC: Cuidado Nutricional nas Doenças do Sistema Nervoso. In:Mahan LK, Stump SE. In Alimentos, Nutrição e Dietoterapia. 9th edition. São Paulo: Roca; 1998:896-7.
10. Borges VC: Repercussão Nutricional nas Doenças Neurológicas da Velhice: Alzheimer e Parkinson. [texto na internet]. [http://www.nutritotal.com.br] webcite Nutrição Total São Paulo. Disponível em; 2003.
11. Chumlea WC, Roche AF, Steinbaugh ML: Estimating stature from knee height for persons 60 to 90 years of age. J Am Geriatric Soc 1985, (33):116-120.
12. Folstein MF, Folstein SE, McHugh PR: Mini-mental state. A practical method for grading the cognitive state of patients for the clinician. J Psychiatr Res 1975, 12:189-198.
13. Bertolucci PHF, Brucki SMD, Campacci SR, Juliano Y: O mini-exame do Estado Mental em uma população geral. Impacto da escolaridade. Arq Neuropsiq 1994, 52:1-7.

14. Hughes CP, Berg L, Danziger WL, Coben LA, Martin RL: A new clinical scale for the stating of dementia. British Journal of Psychiatry 1982, 140:556-572.
15. Morris JC: The Clinical Dementia Rating (CDR): current version and scoring rules. Neurology 1993, 43:2412-2414.
16. Caovilla VP, Canineu PR: Você não está sozinho. ABRAZ. São Paulo: ABRAZ; 2002.
17. Muñoz AM, Agudelo GM, Lopera FJ: Diagnóstico del estado nutricional de los pacientes con demencia tipo Alzheimer registrados en el Grupo de Neurocienciais, Medellín, 2004. Biomédica 2006, 1(26):113-125.
18. Boog MCF: Educação nutricional em serviços públicos de saúde. Cad Saúde Pública 1999, 2(15):S139-S147.
19. Cervato AM, Dernti AM, Latorre MRDO, Marucci MFN: Educação nutricional para adultos: uma experiência positiva em Universidade Aberta para a Terceira Idade. Rev Nutr 2005, 18(1):41-52.
20. Carver AD, Dobson AM: Effects of dietary supplementation of elderly demented hospital residents. J hum Nutr Diet 1995, (8):389-94.
21. Faxen-Irving G, Andren-Olsson B, Geirjerstam AF, Basun H, Cederholm T: The effect of nutritional intervention in elderly subjects residing in group-living for the demented. Eur Jour of clinical 2002, 56(3):221-27.
22. Trelis JJB, López IF: La Alimentación del enfermo de Alzheimer en el âmbito familiar. Nutr Hosp 2004, (19):154-59.
23. Lauque S, Battandier FA, Mansourian R, Guigoz Y, Paintin M: Nourhashemi F, et al. Protein-energy oral supplementation in malnourished nursing-home residents. A controlled trial. Age and Ageing 2000, (29):51-6.
24. Volkert D, Hubsch S, Oster P: Nutritional support and functional status in undernourished geriatric patients during hospitalization and 6 months follow-up. Aging Clin Exp Res 1996, (8):386-95.
25. Norregaard O, Tottrup A, Saaek A: Effects of oral nutritional supplements on adults with chronic obstructive pulmonary disease. Clin Resp Physiol 1987, (23):388.
26. Sampaio ARD, Mannarino IC: Medidas bioquímicas de avaliação do estado nutricional. In Avaliação Nutricional: Aspectos Clínicos e Laboratoriais. Edited by Duarte ACG. São Paulo: Atheneu; 2007:69-73.
27. Roebothan BV, Chandra RK: Relationship between Nutritional Status and Immune Function of Elderly People. Age Ageing 1994, (23):49-53.
28. Spindler AA, Renvall MJ, Nichols JF, Ramsdell JW: Nutritional Status of Patients with Alzheimer's Disease: A 1-Year Study. J AM Diet Assoc 1996, 10(96):1013-1018.
29. Fonseca AM, Soares E: Interdisciplinaridade em grupos de apoio a Familiares e Cuidadores do Portador da Doença de Alzheimer. Rev Saúde Com 2007, 3(1):3-11.

PART III

VITAMINS AND MINERALS

CHAPTER 6

Pharmacokinetics of a Single Oral Dose of Vitamin D3 (70,000 IU) in Pregnant and Non-Pregnant Women

DANIEL E. ROTH, ABDULLAH A. L. MAHMUD,
RUBHANA RAQIB, ROBERT E. BLACK,
AND ABDULLAH H. BAQUI

6.1 BACKGROUND

Vitamin D is essential for the growth and development of the human skeleton throughout the life cycle [1]. There is considerable speculation regarding the potential effects of vitamin D on both skeletal and extra-skeletal aspects of reproductive physiology and fetal development, yet it remains unknown whether there are benefits to improving maternal ante-natal vitamin D status beyond the correction of severe deficiency [2,3]. Clinical trials employing vitamin D dose regimens that safely optimize maternal-fetal vitamin D status will enable testing of these hypotheses [4]. However, very few studies have rigorously addressed vitamin D supple-mentation during pregnancy, and the single-dose vitamin D3 pregnancy trials published to date have provided little insight into pharmacokinetics

Pharmacokinetics of a Single Oral Dose of Vitamin D3 (70,000 IU) in Pregnant and Non-Pregnant Women. © Roth DE, Al Mahmud A, Raqib R, Black RE, and Baqui AH. Nutrition Journal *11,114 (2012), doi:10.1186/1475-2891-11-114. Licensed under Creative Commons Attribution 2.0 Generic License, http://creativecommons.org/licenses/by/2.0/.*

or safety [5,6]. Moreover, there is a near complete absence of pharmaco-logical data in South Asia, where the vitamin D status of pregnant women [7] and young infants [8] is poor in spite of the tropical climate.

The pharmacokinetics of oral vitamin D3 are conventionally de-scribed with respect to its effect on the serum concentration of the pre-dominant circulating metabolite, 25-hydroxyvitamin D ([25(OH)D]), which is a well-established biomarker of systemic vitamin D status [9]. The present study was conducted to assess changes in serum [25(OH)D] and calcium following a single oral vitamin D3 dose (70,000 IU) in non-pregnant women and pregnant women in the third trimester of preg-nancy in Dhaka, Bangladesh. The aim was to generate preliminary phar-macokinetic (PK) and safety data to inform the design of supplementa-tion regimens for use in future larger-scale trials of antenatal vitamin D supplementation in Bangladesh.

6.2 METHODS

6.2.1 PARTICIPANTS

Pregnant and non-pregnant women were enrolled at a clinic in Dhaka, Bangladesh (24°N) from July 2009 to February 2010 if they were aged 18 to <35 years, held permanent residence in Dhaka at a fixed address, and planned to stay in Dhaka for at least four months (Figure 1). Reasons for exclusion were a known medical condition, self-reported current use of any dietary supplements containing vitamin D, use of anti-con-vulsant or anti-mycobacterial medications, severe anemia (hemoglobin concentration <70 g/L), or hypertension at enrollment (systolic blood pressure ≥140 mmHg or diastolic blood pressure ≥90 mmHg on at least two measurements). Pregnant women were excluded if they had major risk factors for preterm delivery (e.g., preterm labor or previous preterm delivery), pregnancy complications or had previously delivered an in-fant with a congenital anomaly or perinatal death. Non-pregnant women were excluded if they were possibly pregnant (e.g., missed recent men-ses) or lactating.

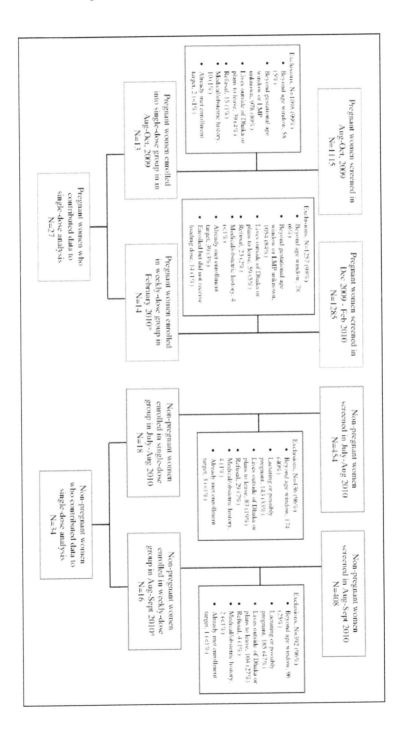

FIGURE 1: Flow diagram of participant screening, exclusions, and enrollment.

All participants in this study received a dose of vitamin D (70,000 IU) at baseline. Primary PK analyses involved participants who did not receive any additional vitamin D throughout follow-up ("single-dose group"). However, to enhance the assessment of 25(OH)D response and safety during the first week of follow-up, an additional cohort of participants who continued to receive weekly vitamin D doses beginning on day 7 ("weekly-dose group") contributed biochemical data to the present analysis for the first 7 days after the 70,000 IU dose (i.e., up to the time preceding their 2nd dose). Findings related to the effect of weekly dosing will be reported elsewhere. Participants were enrolled in stages according to a design that enabled interim analyses and the testing of supplementation regimens in non-pregnant participants prior to their initiation in pregnant women: non-pregnant participants were enrolled in the summer (July to September 2009); pregnant women who received only a single dose were enrolled during the 30th week of gestation in August-September 2009; and, pregnant participants who received the initial dose followed by weekly doses were enrolled at 27 to <31 completed weeks of gestation in February 2010. The study was reviewed and approved by the Institutional Review Board at The Johns Hopkins Bloomberg School of Public Health and the International Center for Diarrheal Disease Research, Bangladesh (ICDDR,B). All participants gave signed informed consent prior to participation. The trial was registered at ClinicalTrials.gov (NCT00938600).

6.2.2 INTERVENTION

Vitamin D3 (cholecalciferol) 70,000 IU (1.75 mg) was administered directly by study personnel. The dose was selected to be intermediate between the doses previously studied in the only two rigorous single-dose vitamin D3 pharmacokinetic studies published at the time our study was designed (50,000 [10] and 100,000 IU [11]), thus providing reassurance in terms of probable safety as well as enabling coherent between-study comparisons. The vitamin D3 supplement (Vigantol Oil, Merck KGaA, Germany) was a liquid formulation (20,000 IU D3/mL). The batch of Vigantol Oil used in the study had a concentration of 20,697 IU/mL according to the manufacturer's certificate of analysis (May, 2009). The stability of the vitamin D3 was established by independent testing of unused Vigantol

Oil at the end of the study (June 2010) in the laboratory of Dr. Reinhold Vieth [12], which revealed a concentration of 19,300 IU/mL (96.5% of the labeled concentration). Participants were advised not to take other vitamin D-containing supplements during the study period, but were permitted to take other micronutrient supplements (including calcium). All pregnant participants were provided standard iron and folic acid supplementation.

6.2.3 FOLLOW-UP

Study personnel assessed participants at least weekly. Non-pregnant participants who received only the single dose participated in weekly follow-up for 10 weeks; pregnant women in the single-dose group were assessed at least weekly until delivery, and then at least three times between delivery and discharge from the study at one-month post-partum. Visits involved a checklist of symptoms related to hypo- and hypercalcemia (decreased appetite, weight loss, vomiting, fever or chills, constipation, abdominal pain, excessive thirst, frequent urination, muscle weakness, back, arm, or leg pain, confusion, or depression), blood pressure measurement, and confirmation of fetal viability.

Abnormal urinalyses, hypertension, reported severe symptoms, or persistence of any mild symptomatic complaints (i.e., decreased appetite, weight loss, vomiting, fever or chills, constipation, abdominal pain, excessive thirst, frequent urination, muscle weakness, back, arm, or leg pain, confusion, or depression) for two consecutive visits prompted referral to the study physician for further evaluation. Participants were referred to an antenatal care physician at the maternity clinic for treatment of urinary tract infections, hypertension, or other medical problems that arose. Participants with obstetric complications were transported to a local tertiary-care hospital with advanced neonatal care facilities. All costs of medical and obstetric care were borne by the study.

6.2.4 SPECIMEN COLLECTION AND BIOCHEMICAL ANALYSES

Participants provided up to six scheduled blood specimens and at least seven urine samples during the 10-week follow-up period beginning on the day of supplement administration (Figure 2). To limit the burden of specimen collection on each individual, yet still enable robust group-level pharmacokinetic and safety analyses, participants were assigned to one of two sampling schedules (A or B) to enhance coverage of the follow-up period (Figure 2). During the first week, specimens were collected at baseline and then additionally on either day 2 or 4 to monitor for possible early transient elevations in serum calcium and to minimize the chance of missing a possible early peak in [25(OH)D]. In the single-dose only groups, blood collection thereafter was scheduled predominantly in the first month because this was when the peak [25(OH)D] [11] and the highest risk of hypercalcemia were anticipated. In pregnant women in the single-dose only group, the specimen collection schedule was continued in the postpartum period if delivery occurred prior to 10 weeks from enrollment. In pregnant participants, venous cord blood samples were collected immediately following delivery of the placenta.

Serum samples (separated from maternal venous and umbilical vein blood) and random spot urine specimens were maintained at +4°C prior to same-day transfer to the laboratory. Sera were frozen at −20 °C. Aliquots for the 25(OH)D assay were shipped at ambient temperature from Dhaka to Toronto (25(OH)D is stable under a range of conditions). Total serum [25(OH)D] was measured with the Diasorin Liaison Total assay in the laboratory of Dr. Reinhold Vieth (Mount Sinai Hospital, Toronto) according to a method previously described [13]. This laboratory participates in and meets the performance targets of the International Vitamin D External Quality Assessment Scheme [14]. Mean within-run coefficient of variation (CV%) was 7.8% (5.8% for specimens with values < 150 nmol/L) and mean between-run CV% was 10.5% (9.0% for specimens <150 nmol/L). Ancillary serum and urine biochemical tests were performed using the AU640 Olympus Autoanalyzer (Olympus Corporation, Japan) at ICDDR,B.

The primary pharmacokinetic (PK) outcome measure was the serum [25(OH)D]; incremental changes from baseline (Δ[25(OH)D]) were calculated as an individual's absolute [25(OH)D] at each visit minus her baseline [25(OH)D]. The primary safety-related outcome was maternal albumin-adjusted serum calcium concentration ([Ca]), calculated using a conventional formula: [Ca]+(0.02*(40-albumin)). The reference range for albumin-adjusted serum calcium was set at $2.10 - 2.60$ mmol/L, the upper limit of which was a conservative threshold relative to those used by: the local laboratory in Dhaka (2.62 mmol/L), the US Institute of Medicine (IOM) 1997 dietary reference intakes (DRIs) for vitamin D (2.75 mmol/L) [15], and the IOM revised 2011 vitamin D DRIs (2.63 mmol/L) [1]. An albumin-adjusted serum calcium concentration >2.60 mmol/L prompted a repeat measurement on a new specimen as soon as possible. Confirmed hypercalcemia was a priori defined as albumin-adjusted serum calcium concentration > 2.60 mmol/L on both specimens, since hypercalcemia caused by vitamin D intoxication would not be expected to resolve within a few days without intervention.

The urinary calcium:creatinine ratio (ca:cr) was expressed as mmol Ca/mmol Cr, and 1.0 was considered the nominal upper limit of the reference range [16]. Any episode of urinary ca:cr>1.0 mmol/mmol prompted a repeat urine ca:cr measurement within one week. In addition, a ca:cr > 0.85 mmol/mmol that was also a 2-fold or greater increase over the lowest previously observed ratio in the same participant prompted repeat urine assessment. Persistent hypercalciuria was defined as ca:cr > 1.0 mmol/ mmol on two consecutive tests, or on two non-consecutive measurements that occurred in the presence of persistent symptoms suggestive of possible hypercalcemia. Persistent hypercalciuria or persistent ca:cr > 0.85 mmol/mmol (under the conditions listed above) were indications for un-scheduled measurement of serum calcium.

6.2.5 STATISTICAL ANALYSES

Continuous outcome variables were described by means, standard deviations (SD), and 95% confidence intervals (95% CI). Non-normally distributed variables (including [25(OH)D]) were described by geometric means with 95% CI's, medians and interquartile ranges (IQR), and were log-transformed for modeling. In the primary PK analysis, the following model-independent PK parameters were estimated for each individual in the single-dose only groups (N=31): 1) maximum observed [25(OH)D] (Cmax); 2) maximum observed Δ[25(OH)D] above baseline (ΔCmax); 3) timing of Cmax in days (Tmax); and 4) area under the Δ[25(OH)D]-time curve (AUC), which was interpreted as a global measure of vitamin D3 bioavailability. Individual participants' AUCs were estimated manually by the trapezoidal method, and negative Δ[25(OH)D] values were zeroed so that the AUC represented the positive area above baseline. AUC was estimated for the first month to enable comparisons to other published PK studies [10,11]. $AUC_{28/35}$ was calculated for either 0 to 28 days or 0 to 35 days, depending on the timing of the blood sampling (Figure 2); similarly, $AUC_{56/70}$ was calculated for the period 0 to 56 days or 0 to 70 days. An individual's average Δ[25(OH)D] during the first 28 days (ΔCavg28) was calculated by dividing AUC_{28} by 28; for between-study comparisons, this measure was expressed per 40,000 IU (1 mg) vitamin D3 by dividing ΔCavg28 by the dose administered (1.75 mg). Cmax, ΔCmax, Tmax, $AUC_{28/35}$, and $AUC_{56/70}$ were summarized within groups by geometric means and 95% CIs, and then log-transformed for one-way analyses of variance (ANOVA) to test for differences between the pregnant and non-pregnant groups. To plot the longitudinal change in [25(OH)D] over time using all available data (N=61), mean [25(OH)D] at each visit were predicted from a linear regression model using a random intercept for each participant, with each visit represented by its own fixed indicator variable. Cross-sectional differences in Δ[25(OH)D] between pregnant and non-pregnant groups at specific days of follow-up were compared by ANOVA. Changes in biochemical ([Ca] and Ca:Cr) and clinical outcomes from baseline were analyzed using generalized estimating equations (GEE) to account for repeated measures. The association between cord venous [25(OH)D] and the corresponding maternal [25(OH)D] closest in time to delivery was analyzed using Pearson correlation.

The target sample size of at least 12 analyzable participants per single-dose group was originally justified as follows: assuming two samples per subject (baseline and peak), a standard deviation for the ΔCmax of 20 nmol/L and an intra-subject correlation of 0.6, we anticipated that at least 12 women in each group would enable the estimation of the mean ΔCmax with 95% confidence bounds of ±10 nmol/L. In all analyses, P values less than 0.05 were considered to be statistically significant, with corrections for multiple comparisons using the Holm method, applied where appropriate [17]. Analyses were conducted using Stata version 10 and 11 (Stata Corporation, College Station, Texas).

		Day														
		0	2	4	7	14	21	28	35	42	49	56	63	65	67	70
Single-dose groups	A	O■	O	O	O	■	O	■	O	■	O	■				■
	B	O■	O	■	O	■	O	■	O	■	O	■	■			O
Weekly-dose groups	C	O■	O	O	O											
	D	O■	O	O	■											

■ Urine collection   O Blood collection

FIGURE 2: Blood and urine specimen collection schedules. Participants in "single-dose only" groups were randomized to one of two schedules (A or B) of specimen collection over a period 70 days. Participants in the "weekly dose" groups were similarly randomized to one of two schedules (C or D); however, the analysis of single-dose pharmacokinetics only included those specimens collected up to and including day 7, preceding administration of the 2nd vitamin D dose.

6.3 RESULTS

In the single-dose only groups, follow-up for the full 10 weeks was completed in all non-pregnant (N=18) and pregnant (N=13) participants; however, the terminal serum sample for one non-pregnant participant (at day 56) was not suitable for analysis. Cord blood specimens were available for 12 of 13 pregnant participants in the single-dose only group. An additional 16 non-pregnant and 14 pregnant participants enrolled in weekly-dose groups contributed at least one [25(OH)D] value on or prior to day 7.

At baseline, pregnant participants had lower average [25(OH)D] than non-pregnant participants (Table 1); this was partly attributable to the design of the study, whereby some pregnant women were enrolled in the winter and all non-pregnant women were enrolled in the summer and fall (Table 2). Pregnant participants were generally younger, more likely to be married, and of a slightly lower socioeconomic status than non-pregnant participants (Table 2).

TABLE 1: Changes in [25(OH)D] following a single dose of 70,000 IU vitamin D3 in non-pregnant and pregnant women in Dhaka, Bangladesha

	All participants	Non-pregnant	Pregnant	Pb
N (all participants)	61	34	27	
Baseline [25(OH)D] (N=61)				
Mean [95% CI]	47 [42, 52]	54 [47, 62]	39 [34, 45]	0.010
Range	21, 96	27, 96	21, 95	
Δ[25(OH)D], Mean [95% CI], nmol/L				
Day 2 (N=27)	20 [15,25]	24 [17, 33]	15 [11,22]	0.037
Day 4 (N=27)	23 [18, 30]	24 [17, 34]	23 [16, 32]	0.800
Day 7 (N=29)	26 [21, 33]	25 [18, 34]	28 [20, 40]	0.134
Day 21 (N=14)	25 [18, 35]	21 [14, 33]	32 [20, 51]	0.003
Day 56 (N=12)	16 [11,23]	14 [9,21]	20 [12, 33]	0.101
Participants in single-dose only groups				
N (% followed more than 7 days)	31 (51%)	18 (53%)	13 (48%)	
# Specimens per participant				
Median	6	6	6	

TABLE 1: *Cont.*

	All participants	Non-pregnant	Pregnant	Pb
Range	3, 6	3, 6	3, 6	
Baseline [25(OH)D] (N=31)				
Mean [95% CI]	48 [41, 56]	52 [42, 64]	43 [34, 55]	0.224
Range	21, 96	27, 96	21, 95	
Tmax, days (N=31)				
Mean [95% CI]	11 [7,18]	9 [4,17]	17 [10,29]	0.134
Range	2, 70	2, 70	2, 70	
Cmax, nmol/L (N=31)				
Mean [95% CI]	85 [77, 93]	87 [75, 101]	82 [72, 92]	0.500
Range	51, 164	51, 164	52, 116	
ΔCmax, nmol/L (N=31)				
Mean [95% CI]	30 [23, 39]	28 [18, 42]	33 [24, 46]	0.486
Range	2, 87	2, 87	9, 52	
Area under the curve, nmol·d/Lc				
$AUC_{56/70}$, Mean (N=30)	935	910	969	0.863
[95% CI]	[651, 1343]	[531, 1559]	[563, 1668]	
$AUC_{28/35}$, Mean (N=31)	591	562	632	0.672
[95% CI]	[448, 780]	[383, 823]	[398, 1003]	
ΔCavg$_{28/35}$ per mg dosed (nmol/L/mg)				
[95% CI]	12 [10,15]	12 [8,15]	14 [10,18]	0.370

a. [25(OH)D] summary measures are geometric means with exponentiated 95% confidence intervals, unless otherwise indicated. b. One-way analysis of variance (ANOVA) test for difference between non-pregnant and pregnant groups. c. AUC was only estimated using data from participants with follow-up to the end of the interval of interest (i.e., 28/35 days or 56/70 days). d. Average [25(OH)D] over the first 28 or 35 days, per mg of the single vitamin D3 dose. Arithmetic means and 95% confidence intervals reported because these estimates had an approximately normal distribution.

6.3.1 PHARMACOKINETIC OUTCOMES

There was substantial inter-individual variation in the shape and magnitude of 25(OH)D responses to a single oral dose of 70,000 IU vitamin D3. However, the population-average pattern consisted of an abrupt increase in [25(OH)D] in the first week, followed by a peak within the first three weeks, and then a gradual return to baseline over the ensuing two months in both non-pregnant and pregnant participants (Figure 3). The average [25(OH)D] remained marginally above baseline at ten weeks after supplementation. There were minor differences between the pregnant and non-pregnant groups in the average Δ[25(OH)D] throughout follow-up (Table 1). In particular, [25(OH)D] rose more rapidly and the peak average occurred earlier in the non-pregnant group. This was demonstrated by the significantly greater Δ[25(OH)D] on day 2, the significantly lower Δ[25(OH)D] on day 21, and the slightly earlier occurrence of Tmax in non-pregnant vs. pregnant women (Table 1). Moreover, there was greater variance in the early Δ[25(OH)D] in non-pregnant vs. pregnant participants (Figure 3). The highest [25(OH)D] in any non-pregnant participant was 164 nmol/L, whereas the maximum in any pregnant participant was 116 nmol/L. On average, pregnant women had slightly lower absolute Cmax, but the mean maximal rise in [25(OH)D] (i.e., ΔCmax) and AUC were similar in pregnant and non-pregnant women (Table 1). Overall, the [25(OH)D] was an average of 19 nmol/L (95% CI, 14 to 25) higher than baseline during the first month after supplementation, which corresponded to a gain of approximately 12 nmol/L per mg of the vitamin D3 dose (Table 1).

TABLE 2: Personal and household characteristics of participants at enrollment

	Single-dose only group			All participants		
	Non-pregnant	Pregnant	P	Non-pregnant	Pregnant	Pa
# Enrolled	18	13		34	27	
Month of enrollment						
July-August, 2009	18 (100 %)	5 (38 %)	<0.001	33 (97%)	5 (19%)	<0.001
Sept-Oct 2009	0	8 (62 %)		1 (3%)	8 (30%)	
February 2010	0	0		0	14 (52%)	
Age (years), Mean (±SD)	23.9 (±3.8)	20.9 (±2.7)	0.022	24.2 (±4.1)	21.6 (±2.9)	0.006
Married	11 (61%)	13 (100%)	0.025	23 (68%)	27 (100%)	0.001

TABLE 2: *Cont.*

	Single-dose only group			All participants		
	Non-pregnant	Pregnant	P	Non-pregnant	Pregnant	Pa
Education level attained						
None	1 (6%)	2 (15%)	0.750	3 (9%)	6 (22%)	0.293
Primary	11 (61%)	7 (54%)		21 (62%)	16 (59%)	
Secondary or higher	6 (33%)	4 (31%)		10 (29%)	5 (19%)	
Husband's education level						
None	2 (18%)	3 (23%)	1.000	2 (9%)	4 (15%)	0.786
Primary	4 (36%)	4 (31%)		10 (43%)	13 (48%)	
Secondary or higher	5 (45%)	6 (46%)		11 (48%)	10 (37%)	
Home ownership	6 (33%)	1 (8%)	0.191	7 (21%)	2 (7%)	0.276
House constructed from cement, brick or tileb						
Floor	18 (100%)	11 (85%)	0.168	33 (98%)	22 (81%)	0.079
Walls	16 (89%)	10 (77%)	0.625	30 (88%)	18 (67%)	0.042
Roof	6 (33%)	6 (46%)	0.710	13 (38%)	7 (26%)	0.412
Height (cm), mean (±SD)	149.7 (±3.7)	150.3 (±3.9)	0.685	150.8 (±4.3)	150.5 (±4.3)	0.758

a. ANOVA for comparisons of continuous variables, Fisher's exact test for categorical variables. b. In comparison to tin or natural materials (e.g., earth, bamboo).

6.3.2 SAFETY OUTCOMES

The supplement was tasteless and well tolerated and there were no supplement-related adverse events (Table 3). The stillbirth and newborn deaths were explained by medical problems, and there was no evidence that either was related to the vitamin D supplementation, given their timing (i.e., did not occur at peak [25(OH)D]) and the absence of biochemical evidence of vitamin D toxicity in the mother (Table 3). Postmortem examinations were not feasible in the study setting. Two other AEs resolved without complications and occurred in the absence of evidence of vitamin D toxicity (Table 3). Pregnancy and birth outcome metrics were consistent with expectations for the source population (Table 4).

TABLE 4: Pregnancy and newborn outcomes for pregnant participants who received only a single dose of 70,000 IU vitamin D at enrollment and were followed up to delivery

N	13
Gestational age at birth, weeks (by LMP) a Mean (±SD)	38.8 (±1.8)
Range	35.7 – 42.0
Preterm, n (%)	2 (15%)
Birth weightb (g)	
Mean (±SD) c	2441 (±354)
Range (g)	1890 – 3005
n/N (%) Low Birth Weight	6/12 (50%)
Delivery mode, n/N (%) Cesarean section d	8/13 (62%)
Sex, n (%) female	5 (38%)
Live birthse	12/13
Alive at 1 month of agef	11/13

a. In a sample of 113 deliveries at the study site (October 2009 to January 2010) for which there was a recalled first day of last menstrual period, the mean gestational age at birth was estimated to be 39.7 weeks (±2.2). b. Only includes the 12 liveborn infants. c. In a consecutive sample of 362 liveborn infants delivered at the study site (October 2009 to January 2010), the mean birth weight was 2780 g (±440). d. In a consecutive sample of 369 deliveries at the study site (October 2009 to January 2010), there were 199 cesarean deliveries (54%). e. There was one stillbirth. In a consecutive sample of 369 deliveries at the study site (October 2009 to January 2010), there were 7 stillbirths (2%). f. There was one neonatal death at 3 days of age.

Changes in average serum calcium concentrations (Figure 4) and urinary calcium excretion (Figure 5) occurred during the early phase of [25(OH)D] escalation. In non-pregnant participants, a transient increase in albumin-adjusted serum [Ca] from baseline was notable on day 4 (Table 5;

Figure 4). The corresponding change in unadjusted total serum [Ca] was smaller and non-significant, and the raised adjusted [Ca] coincided with a lower average serum albumin on day 4 (difference versus baseline, -1.23 g/L; 95% CI, -2.12 to −0.34). In pregnant participants, there was an initial increase in albumin-adjusted [Ca] beginning on day 2 that persisted until nearly the end of the observation period (Figure 4), but the difference from baseline was only statistically significant on day 7 (Table 6). The unadjusted total [Ca] did not vary greatly from baseline and serum albumin remained relatively stable until the end of the 70-day follow-up, when many of the participants were post-partum.

There were no episodes of confirmed hypercalcemia according to the study definition, and no isolated albumin-adjusted [Ca] values greater than the recent IOM upper limit of normal of 2.63 mmol/L. One pregnant participant had a single albumin-adjusted [Ca] = 2.61 mmol/L at one-week postpartum (70 days after dose administration) corresponding to a normal total [Ca] (2.51 mmol/L; serum albumin concentration was 35.8 g/L) that was within the reference range on repeat testing 4 days later (Table 3; Figure 6). A further follow-up one week following the first abnormal result was also normal (albumin-adjusted serum [Ca] of 2.44 mmol/L). This participant also had two non-consecutive episodes of urinary ca:cr higher than 1.0 mmol/mmol during follow-up (Figure 6). Her serum biochemical patterns were consistent with the expected changes in the perinatal period, including a gradual increase in albumin-adjusted serum [Ca] towards the end of the antenatal period and a rapid increase in serum albumin in the post-partum period [18]. Furthermore, there was no temporal association between the rise in [25(OH)D] and either the occurrence of isolated peaks in urine ca:cr or the isolated elevated [Ca] (Figure 6).

None of the participants manifested persistent hypercalciuria according to the study definition, or using a more conservative threshold of 0.85 mmol/mmol. In non-pregnant participants, the Ca:Cr increased from baseline but differences were only statistically significant at day 7 and 14 (Table 5). In pregnant participants, the increases in average Ca:Cr above baseline were more persistent and were statistically significant on all days except day 42, 49, and 70 (Table 6). There was no overall difference in the average ca:cr between non-pregnant and pregnant participants (P= 0.857).

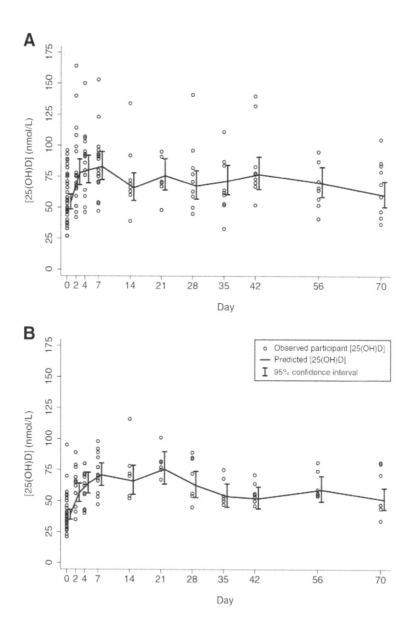

FIGURE 3: Serum [25(OH)D] in non-pregnant (A) and pregnant (B) participants following administration of 70,000 IU vitamin D3 at day 0. Predicted mean [25(OH)D] and 95% confidence intervals were estimated in a random-intercept regression model of ln[25(OH)D] as a function of time.

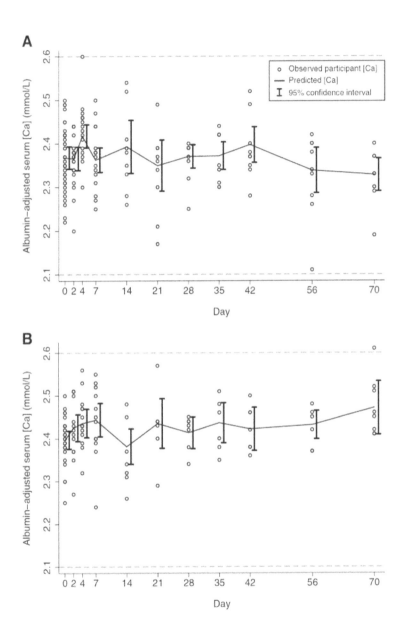

FIGURE 4: Albumin-adjusted serum calcium concentration ([Ca]) in non-pregnant (A) and pregnant participants (B) following administration of vitamin D3 70,000 IU at day 0. Dashed horizontal lines represent upper and lower bounds of the reference range. Predicted means and 95% confidence intervals were estimated in a linear regression model using GEE.

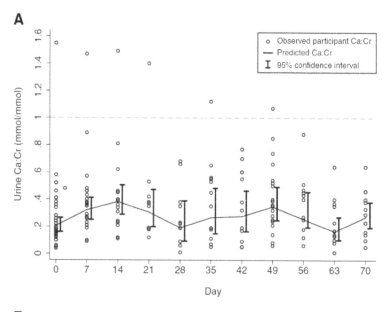

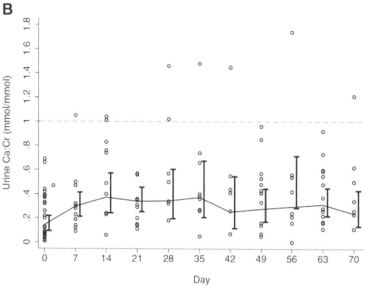

FIGURE 5: Calcium:creatinine ratios (Ca:Cr) in spot urine specimens from non-pregnant (A) and pregnant participants (B) following administration of vitamin D3 70,000 IU at day 0. Predicted means and 95% confidence intervals were estimated in a linear regression model using GEE, in which log-transformed Ca:Cr was modeled as a function of time.

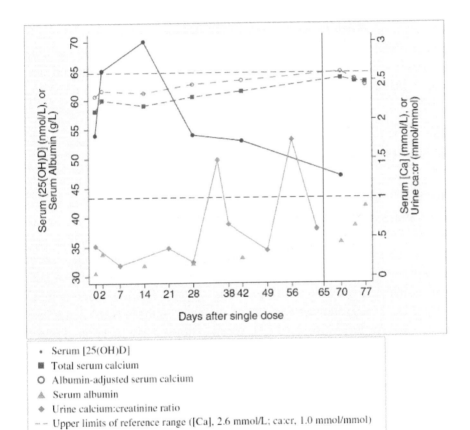

FIGURE 6: Serum and urine biochemistry in a pregnant participant with two episodes of urine ca:cr > 1.0 mmol/mmol and one episode of serum albumin-adjusted [Ca] > 2.60 mmol/L. Vertical line indicates timing of delivery at 39 weeks gestation.

6.3.3 CORD BLOOD BIOCHEMISTRY

Among participants who had received a single dose at baseline and for whom cord blood specimens were collected (N=12), the geometric mean cord serum [25(OH)D] was 50 nmol/L (95% CI, 40 to 62; range, 29 to 80). All cord serum albumin-adjusted [Ca] were within the normal range. The cord ln[25(OH)D] was moderately correlated with the maternal ln[25(OH)D] closest to the time of delivery (Pearson rho=0.64, P=0.02), and the average ratio of cord:maternal [25(OH)D] (N=12) was 0.88 (95% CI, 0.76 –1.02).

6.4 DISCUSSION

This single-dose oral vitamin D3 pharmacokinetic study generated novel observations regarding the biochemical response to vitamin D3 in women of reproductive age in South Asia. Overall, we found that the average response was similar to that reported for non-pregnant adults in other geographic settings. The occurrence of the maximal mean [25(OH)D] in the first month was consistent with previous studies of single-dose vitamin D3 (1.25 to 15 mg) administered to non-pregnant adults in North America, Europe and Australia [10,11,19-23]. When expressed as a function of vitamin D3 dose (assuming the rise is linearly proportional to dose), the mean overall ΔCmax of 30 nmol/L (28 nmol/L in non-pregnant and 33 nmol/L in pregnant participants) represented an average maximal rise in [25(OH)D] of ~17 nmol/L per mg D3. This estimate was similar to those of previous studies from which relevant inferences could be drawn, in which the average ΔCmax ranged from 12 to 16 nmol/L per mg of vitamin D3 [10,11,19,20].

We are not aware of previous single-dose vitamin D3 pharmacokinetic studies in pregnancy to which the present findings can be directly compared. However, there are emerging data regarding the efficacy and safety of high-dose continuous regimens in pregnancy; for example, Hollis et al. reported that 4000 IU/day vitamin D3 initiated in the 2nd trimester yielded an increase in mean [25(OH)D] from 58 nmol/L to 111 nmol/L at delivery among women in South Carolina, without inducing hypercalce-mia or other observed adverse effects [24]. In comparison, Vieth observed in non-pregnant adults that 4000 IU/day led to an increase in mean [25(OH)D] from 38 to 96 nmol/L at steady-state[16]. Thus, from a pharmacokinetic standpoint, the Hollis et al. findings are in accord with our conclusion that pregnancy does not substantially alter the 25(OH)D response to vitamin D3.

There was substantial inter-individual variability in 25(OH)D respons-es. Many participants demonstrated a rapid rise in [25(OH)D] during the first week, which is similar to the response to an acute dose of ultraviolet radiation exposure [25]; but distinct from the more gradual effects of other forms of exogenous vitamin D intake (e.g., oral D2 ingestion [10]). Sev-eral non-pregnant participants demonstrated peak [25(OH)D] as early as two days after supplement delivery, and there was notably wider variabil-ity in responses in the group of non-pregnant participants during the early escalation phase compared to pregnant participants. It is possible that the greater apparent variability was an artifact due to lower precision of the 25(OH)D assay at higher [25(OH)D], given the higher average [25(OH)D] in non-pregnant women. Higher concentrations of vitamin D-binding protein during pregnancy [[26] may have efficiently buffered the absorbed vitamin D3 and slowed its transport to the liver where it undergoes 25-hy-droxylation [27]].

Vitamin D3 bioavailability (measured by mean AUC and dose-adjust-ed $\Delta Cavg_{28}$) differed minimally between the non-pregnant and pregnant groups, and between-group differences were overshadowed by between-subject variability. The overall $\Delta Cavg_{28}$ (i.e., estimated average [25(OH)D] rise from baseline in the first month, expressed per milligram of vita-min D3) was 12 nmol/L/mg based on an aggregate analysis of individual empiric AUCs. This result was the same as the $\Delta Cavg_{28}$ of ~12 nmol/L/mg found in studies of non-pregnant adults using 50,000 IU and 100,000 IU [11]], and similar to an extrapolated estimate of 13 nmol/L/mg based on data reported for a single dose of 300,000 IU in elderly adults [19]. The $\Delta Cavg_{28}$ provides a useful summary measure for between-study compari-sons because most of a single ingested vitamin D3 dose is converted to 25(OH)D within one month [11]. The consistency of the present findings with $\Delta Cavg_{28}$ estimates from previous studies supports the contention by Heaney et al. that 25(OH)D bioavailability is proportional to vitamin D3 input across a wide dose range (1.25 to 7.5 mg) [11]. Notably, $\Delta Cavg_{28}$ extracted from a study by Cipriani et al. was somewhat lower (~ 8 nmol/L/mg) [20]. We speculate that the massive dose administered in that study (600,000 IU) saturated the hepatic 25-hydroxylase system, resulting in the engagement of subsidiary vitamin D catabolic pathways which reduced the 25(OH)D yield.

The single vitamin D3 dose of 70,000 IU did not provoke hypercalcemia or hypercalciuria in non-pregnant or pregnant participants, and available data indicated that adverse perinatal events were neither temporally nor mechanistically linked to vitamin D supplementation. An isolated serum [Ca] value above the reference range in one pregnant participant occurred in the early post-partum period, when albumin-adjusted [Ca] typically peaks [18]. This was not due to vitamin D toxicity because her [25(OH)D] at the time was 47 nmol/L and the [Ca] rapidly and spontaneously normalized. However, it is important to acknowledge that there were significant increases in average [Ca] and urine ca:cr. Changes in serum [Ca] were not reportedly significant in studies by Ilahi [11], Armas [10], or Romagnoli [19], but Cipriani et al. demonstrated that the administration of a single dose of 600,000 IU to healthy young adults caused an increase in serum [Ca] at 3 days, coinciding with peak serum concentrations of both 25(OH)D and the active metabolite, 1,25-dihydroxyvitamin D (1,25(OH)2D) [20]. Therefore, upward deflections in the serum and urine biomarkers of calcium homeostasis signaled a need to be cautious about the transient effects of large sudden influxes of vitamin D, and the risk of dose-dependent toxicity.

There were several limitations of this study. First, although we were able to closely monitor the participants to gain preliminary PK and safety data in this population, the small sample size limited the precision of effect estimates and comparisons of non-pregnant and pregnant participants. Moreover, we did not have adequate power to adjust for differences in the baseline characteristics of the pregnant and non-pregnant groups, although we did not expect minor variations in age or socioeconomic status to influence biochemical responses. Second, the low number of scheduled blood specimens collected from each individual compromised the precision of the estimates of individual-level PK parameters. The number was limited by available funds and the expected acceptability of the procedure by participants based on pre-study consultation with local community members. Third, the fixed timing of specimen collection had the disadvantage of leaving gaps in the [25(OH)D]-time curve where no data were available. Fourth, the study lacked an unsupplemented control group. The analysis was challenged by the substantial inter-individual variability in responses to supplementation, which was expected based on previous reports [28]. Several participants had fluctuating [25(OH)D], without a single clear peak and decline, and some manifested seemingly paradoxical responses, with initial declines in [25(OH)D] after D3 ingestion. These erratic patterns could not easily be explained on the basis of known vitamin D pharmacokinetics, but were most likely attributable to small-sample artifacts, biological variability in the absorption and metabolism of vitamin D, and inherent imprecision in the laboratory assessment of [25(OH)D]. Nonetheless, the data yielded coherent population-averaged interpretations that were consistent with published data from non-pregnant adults in other settings.

6.5 CONCLUSIONS

Comparisons of pregnant (third-trimester) to non-pregnant participants, as well as comparisons to previously published PK studies in non-pregnant adults, suggested that the effects of pregnancy on the 25(OH)D response to vitamin D3 were relatively minor and did not substantially impact overall bioavailability. Likewise, we did not document any notable pregnancy-related hypersensitivity to a vitamin D dose of 70,000 IU in terms of its effects on calcium homeostasis. However, the unpredictability of the 25(OH)D response at the individual level, previous reports of adverse effects of large single doses [23], and the theoretical disadvantages of excessive fluctuations in vitamin D status [29] suggest that the use of large single or infrequent intermittent doses of vitamin D3 may be physiologically disadvantageous despite its practical appeal. Therefore, these data principally support the further investigation of single doses equal to or less than 70,000 IU in the context of intermittent (e.g., weekly or biweekly) antenatal dosing regimens.

REFERENCES

1. Ross AC, Taylor CL, Yaktine AL, Del Valle HB: Committee to Review Dietary Reference Intakes for Vitamin D and Calcium, Institute of Medicine. Dietary Reference Intakes for Calcium and Vitamin D. Washington, DC: The National Academies Press; 2010.

2. ACOG Committee Opinion No. 495: Vitamin D: screening and supplementation during pregnancy. Obstet Gynecol 2011, 118:197-198.

3. Specker BL: Does vitamin D during pregnancy impact offspring growth and bone? Proc Nutr Soc 2011, FirstView:1-8.

4. Roth DE: Vitamin D, supplementation during pregnancy: safety considerations in the design and interpretation of clinical trials. J Perinatol 2011, 31:449-459.

5. Sahu M, Das V, Aggarwal A, Rawat V, Saxena P, Bhatia V: Vitamin D replacement in pregnant women in rural north India: a pilot study. Eur J Clin Nutr 2009, 63:1157-1159.

6. Zeghoud F, Garabedian M, Jardel A, Bernard N, Melchior J: Administration of a single dose of 100,000 U.I. of vitamin D3 in the pregnant woman in winter. The effect on blood calcium level of the newborn infant. J Gynecol Obstet Biol Reprod (Paris) 1988, 17:1099-1105.

7. Marwahaa RK, Tandona N, Chopraa S, Agarwala NM, Garga K, Sharmaa B, Kanwara RS, Bhadraa K, Singha S, et al.: Vitamin D status in pregnant Indian women across trimesters and different seasons and its correlation with neonatal serum 25-hydroxyvitamin D levels. Br J Nutr 2011, 106:1383-1389.

8. Roth DE, Shah MR, Black RE, Baqui AH: Vitamin D status of infants in northeastern rural bangladesh: preliminary observations and a review of potential determinants. J Health Popul Nutr 2010, 28:458-469.

9. Seamans KM, Cashman KD: Existing and potentially novel functional markers of vitamin D status: a systematic review. Am J Clin Nutr 2009, 89:1997S-2008S.

10. Armas LA, Hollis BW, Heaney RP: Vitamin D2 is much less effective than vitamin D3 in humans. J Clin Endocrinol Metab 2004, 89:5387-5391.

11. Ilahi M, Armas LA, Heaney RP: Pharmacokinetics of a single, large dose of cholecalciferol. Am J Clin Nutr 2008, 87:688-691.

12. Maalouf J, Nabulsi M, Vieth R, Kimball S, El-Rassi R, Mahfoud Z, El-Hajj Fuleihan G: Short- and long-term safety of weekly high-dose vitamin D3 supplementation in school children. J Clin Endocrinol Metab 2008, 93:2693-2701.

13. Wagner D, Hanwell HE, Vieth R: An evaluation of automated methods for measurement of serum 25-hydroxyvitamin D. Clin Biochem 2009, 42:1549-1556.

14. Carter GD, Berry JL, Gunter E, Jones G, Jones JC, Makin HL, Sufi S, Wheeler MJ: Proficiency testing of 25-hydroxyvitamin D (25-OHD) assays. J Steroid Biochem Mol Biol 2010, 121:176-179.

15. Standing Committee on the Scientific Evaluation of Dietary Reference Intakes: Food and Nutrition Board IOM. In Dietary Reference Intakes for Calcium, Phosphorus, Magnesium, Vitamin D, and Fluoride. Washington: National Academy Press; 1997.

16. Vieth R, Chan PC, MacFarlane GD: Efficacy and safety of vitamin D3 intake exceeding the lowest observed adverse effect level. Am J Clin Nutr 2001, 73:288-294.

17. Gordon AY, Salzman P: Optimality of the Holm procedure among general step-down multiple testing procedures. Stat Probability Lett 2008, 78:1878-1884.

18. Payne RB, Little AJ, Evans RT: Albumin-adjusted calcium concentration in serum increases during normal pregnancy. Clin Chem 1990, 36:142-144.

19. Romagnoli E, Mascia ML, Cipriani C, Fassino V, Mazzei F, D'Erasmo E, Carnevale V, Scillitani A, Minisola S: Short and long-term variations in serum calciotropic hormones after a single very large dose of ergocalciferol (vitamin D2) or cholecalciferol (vitamin D3) in the elderly. J Clin Endocrinol Metab 2008, 93:3015-3020.

20. Cipriani C, Romagnoli E, Scillitani A, Chiodini I, Clerico R, Carnevale V, Mascia ML, Battista C, Viti R, et al.: Effect of a single oral dose of 600,000 IU of cholecalciferol on serum calciotropic hormones in young subjects with vitamin D deficiency: a prospective intervention study. J Clin Endocrinol Metabol 2010, 95:4771-4777.

21. Bacon CJ, Gamble GD, Horne AM, Scott MA, Reid IR: High-dose oral vitamin D3 supplementation in the elderly. Osteoporos Int 2009, 20:1407-1415.

22. Weisman Y, Schen RJ, Eisenberg Z, Amarilio N, Graff E, Edelstein-Singer M, Goldray D, Harell A: Single oral high-dose vitamin D3 prophylaxis in the elderly. J Am Geriatr Soc 1986, 34:515-518.

23. Sanders KM, Stuart AL, Williamson EJ, Simpson JA, Kotowicz MA, Young D, Nicholson GC: Annual high-dose oral vitamin D and falls and fractures in older women: a randomized controlled trial. JAMA 2010, 303:1815-1822.

24. Hollis BW, Johnson D, Hulsey TC, Ebeling M, Wagner CL: Vitamin D supplementation during pregnancy: Double blind, randomized clinical trial of safety and effectiveness. J Bone Miner Res 2011, 26:2341-2357.

25. Vähävihu K, Ylianttila L, Kautiainen H, Viljakainen H, Lamberg-Allardt C, Hasan T, Tuohimaa P, Reunala T, Snellman E: Narrowband ultraviolet B course improves vitamin D balance in women in winter. Br J Dermatol 2010, 162:848-853.

26. Bouillon R, Van Assche FA, Van Baelen H, Heyns W, De Moor P: Influence of the vitamin D-binding protein on the serum concentration of 1,25-dihydroxyvitamin D3. Significance of the free 1,25-dihydroxyvitamin D3 concentration. J Clin Invest 1981, 67:589-596.

27. Haddad JG, Matsuoka LY, Hollis BW, Hu YZ, Wortsman J: Human plasma transport of vitamin D after its endogenous synthesis. J Clin Invest 1993, 91:2552-2555.

28. Heaney RP: Vitamin D, and calcium interactions: functional outcomes. Am J Clin Nutr 2008, 88:541S-544S.

29. Vieth R: How to optimize vitamin D supplementation to prevent cancer, based on cellular adaptation and hydroxylase enzymology. Anticancer Res 2009, 29:3675-3684.

There are several tables that are not available in this version of the article. To view this additional information, please use the citation on the first page of this chapter.

CHAPTER 7

Causal Relationship between Obesity and Vitamin D Status: Bi-Directional Mendelian Randomization Analysis of Multiple Cohorts

KARANI S. VIMALESWARAN, DIANE J. BERRY, CHEN LU, EMMI TIKKANEN, STEFAN PILZ, LINDA T. HIRAKI, JASON D. COOPER, ZARI DASTANI, RUI LI, DENISE K. HOUSTON, ANDREW R. WOOD, KARL MICHAËLSSON, LIESBETH VANDENPUT, LINA ZGAGA, LAURA M. YERGES-ARMSTRONG, MARK I. MCCARTHY, JOSÉE DUPUIS, MARIKA KAAKINEN, MARCUS E. KLEBER, KAREN JAMESON, NIGEL ARDEN, OLLI RAITAKARI, JORMA VIIKARI, KURT K. LOHMAN, LUIGI FERRUCCI, HÅKAN MELHUS, ERIK INGELSSON, LIISA BYBERG, LARS LIND, MATTIAS LORENTZON, VEIKKO SALOMAA, HARRY CAMPBELL, MALCOLM DUNLOP, BRAXTON D. MITCHELL, KARL-HEINZ HERZIG, ANNELI POUTA, ANNA-LIISA HARTIKAINEN, THE GENETIC INVESTIGATION OF ANTHROPOMETRIC TRAITS (GIANT) CONSORTIUM, ELIZABETH A. STREETEN, EVROPI THEODORATOU, ANTTI JULA, NICHOLAS J. WAREHAM, CLAES OHLSSON, TIMOTHY M. FRAYLING, STEPHEN B. KRITCHEVSKY, TIMOTHY D. SPECTOR, J. BRENT RICHARDS, TERHO LEHTIMÄKI, WILLEM H. OUWEHAND, PETER KRAFT, CYRUS COOPER, WINFRIED MÄRZ, CHRIS POWER, RUTH J. F. LOOS, THOMAS J. WANG, MARJO-RIITTA JÄRVELIN, JOHN C. WHITTAKER, AROON D. HINGORANI, AND ELINA HYPPÖNEN

Causal Relationship between Obesity and Vitamin D Status: Bi-Directional Mendelian Randomization Analysis of Multiple Cohorts. © Vimaleswaran KS et al. PLoS MEDICINE *10*,2 (2013), doi:10.1371/ journal.pmed.1001383. Licensed under Creative Commons Attribution License, http://creativecommons.org/licenses/by/2.0/.

7.1 INTRODUCTION

The prevalence of obesity has increased in the last two decades and it is presently the most common and costly nutritional problem [1]–[4]. In the United States, one-third of the population is affected by obesity, according to the National Health and Nutrition Examination Survey [5]. Despite a known genetic contribution, the increase in obesity prevalence has been largely attributed to lifestyle changes, which means that it is amenable to modification through public health and other interventions [6].

Vitamin D deficiency is another increasingly prevalent public health concern in developed countries [7]–[9], and there is evidence that vitamin D metabolism, storage, and action both influence and are influenced by adiposity. Observational studies have reported an increased risk of vitamin D deficiency in those who are obese; however, the underlying explanations and direction of causality are unclear [10]. Active vitamin D (1,25-dihydroxyvitamin D) may influence the mobilisation of free fatty acids from the adipose tissue [11]. In vitro experiments in rats have also shown that large doses of vitamin D2 lead to increases in energy expenditure due to uncoupling of oxidative phosphorylation in adipose tissues [12]. However, randomized controlled trials (RCTs) testing the effect of vitamin D supplementation on weight loss in obese or overweight individuals have provided inconsistent findings [13]–[15]. It has also been suggested that obesity could result from an excessive adaptive winter response, and that the decline in vitamin D skin synthesis due to reduced sunlight exposure contributes to the tendency to increase fat mass during the colder periods of the year [16],[17]. However, vitamin D is stored in the adipose tissue and, hence, perhaps the most likely explanation for the association is that the larger storage capacity for vitamin D in obese individuals leads to lower circulating 25-hydroxyvitamin D [25(OH)D] concentrations, a marker for nutritional status [18].

In the Mendelian randomization (MR) approach, causality is inferred from associations between genetic variants that mimic the influence of a modifiable environmental exposure and the outcome of interest [19]. If lower vitamin D intake/status is causally related to obesity, a genetic variant associated with lower 25(OH)D concentrations should be associated with higher body mass index (BMI) (in proportion to the effect on 25(OH)

D). Conversely, if obesity leads to lower vitamin D status, then genetic variants associated with higher BMI should be related to lower 25(OH)D concentrations. The genetic associations, unlike the directly observed associations for vitamin D intake/status, should be less prone to confounding by lifestyle and socio-economic factors and be free from reverse causation as genotypes are invariant and assigned at random before conception [20]. The use of multiple SNPs to index the intermediate exposure of interest increases power and reduces the risk of alternative biological pathways (pleiotropy) affecting the observed associations between the genotype and the outcome [21],[22].

In the present study, we investigated the relationship between BMI, a commonly used measure for monitoring the prevalence of obesity at the population level, and vitamin D status and we inferred causality by using genetic variants as instruments in bi-directional MR analyses. Meta-analysis included data from 21 studies comprising up to 42,024 individuals.

7.2 METHODS

7.2.1 ETHICS STATEMENT

All participants provided written, informed consent, and ethical permission was granted by the local research ethics committees for all participating studies.

7.2.2 PARTICIPANTS

The collaboration investigating the association of vitamin D and the risk of cardiovascular disease and related traits (D-CarDia) consists of European ancestry cohorts from the United Kingdom (UK), United States (US), Canada, Finland, Germany, and Sweden. This study comprised a meta-analysis of directly genotyped and imputed SNPs from 21 cohorts totalling 42,024 individuals (Table 1). An expanded description of the participating studies is provided in the Text S2.

TABLE 1: Characteristics of the study cohorts stratified by sex.

Study Name	Sample Size*, n (Men/Women)	Men			Women			Combined		
		Age (y) (Mean± SD)	Geometric Mean BMI (kg/m²)	25(OH)D (nmol/l)	Age (y) (Mean± SD)	Geometric Mean BMI (kg/m²)	25(OH)D (nmol/l)	Age (y) (Mean± SD)	Geometric Mean BMI (kg/m²)	25(OH)D (nmol/l)
1958 British Birth cohort (1958BC)	3,711/3,703	45.2 ± 0.4	27.4	53.1	45.2 ± 0.4	26.3	51.2	45.2 ± 0.4	26.9	52.1
1966 North Finland Birth cohort (NFBC1966)	2,192/2,261	31.1 ± 0.40	25.0	63.8	31.1 ± 0.3	23.7	62.7	31.1 ± 0.4	24.3	63.2
Framingham Heart Study (FHS)	2,678/2,978	46.9± 13.0	25.3	68.4	46.4± 13.1	25.8	71.9	46.6± 13.1	26.8	69.7
The Ludwigshafen Risk and Cardiovascular Health study (LURIC)	2,299/999	61.8± 10.7	27.3	39.5	64.7 ± 10.2	26.9	32.6	62.6± 10.6	27.2	37.3
Hertfordshire Cohort Study (HCS)	586/624	64.3 ± 2.6	26.7	44.8	65.7 ± 2.5	26.9	39.0	65.0± 2.6	26.9	41.7
UK Blood Services Common Control Collection (UKBS-CC)	1,310/1,298	45.2 ± 11.8	26.3	50.4	42.3± 12.5	25.5	55.2	43.8± 12.2	26.1	52.5
Young Finns	907/1,077	37.6± 5.1	26.6	54.1	37.6± 5.0	24.8	57.9	37.6± 5.0	25.5	56.3
Canadian Multi-centre Osteoporosis Study (CaMos)	709/ 1,588	62.3 ± 18.2	26.8	64.1	65.2 ± 15.8	26.8	64.1	64.3± 16.6	26.8	64.1

TABLE 1: *Cont.*

Study Name	Sample Size*, n (Men/Women)	Men			Women			Combined		
		Age (y) (Mean± SD)	Geometric Mean BMI (kg/m²)	25(OH)D (nmol/l)	Age (y) (Mean± SD)	Geometric Mean BMI (kg/m²)	25(OH)D (nmol/l)	Age (y) (Mean± SD)	Geometric Mean BMI (kg/m²)	25(OH)D (nmol/l)
Twins UK	176/1,754	51.0± 13.2	26.1	64.1	51.2± 12.9	25.0	68.7	51.2± 13.2	25.3	68.0
Health, Aging and Body Composition Study (Health ABC)	829/729	74.9± 2.9	26.8	68.4	74.7± 2.8	25.5	66.3	74.8± 2.9	26.3	67.7
The Health Professionals Follow-up Study-CHO (HPFS-CHD)	1,245/ -	63.8± 8.6	25.5	56.8	—	—	—	—	—	—
InCHIANTI Study	496/598	67.2 ± 15.4	26.8	50.4	69.1 ± 15.6	26.8	38.1	68.3± 15.5	26.8	43.4
Uppsala Longitudinal Study of Adult Men (ULSAM)	1,194/ -	71.0± 0.6	26.1	65.8	—	—	—	—	—	—
Prospective Investigation of the Vasculature in Uppsala Seniors (PIVUS)	500/499	70.1 ± 0.2	26.8	56.8	70.2 ± 0.1	26.7	70.2 ± 0.2	—		
The Gothenburg Osteoporosis and Obesity Determinants study (GOOD)	921/ -	18.9 ± 0.6	22.1	61.8	—	—	—	—	—	—

TABLE 1: *Cont.*

Study Name	Sample Size*, n (Men/Women)	Men Age (y) (Mean± SD)	Geometric Mean BMI (kg/m²)	25(OH)D (nmol/l)	Women Age (y) (Mean± SD)	Geometric Mean BMI (kg/m²)	25(OH)D (nmol/l)	Combined Age (y) (Mean± SD)	Geometric Mean BMI (kg/m²)	25(OH)D (nmol/l)
Cancer Genetic Markers of Susceptibility, case control study of breast cancer (NHS-CGEMS)	-/870	—	—	—	59.6± 5.8	25.0	74.4	—	—	—
Health2000 GenMets Study (GENMETS)	397/424	49.2 ± 10.4	25.3	44.7	52.0± 11.6	24.8	45.2	50.7 ± 11.1	25.0	44.7
MRC Ely study	323/435	53.8± 7.8	25.8	57.7	53.2± 7.6	25.3	50.1	53.5 ± 7.7	25.5	53.2
Study of Colorectal Cancer in Scotland (SOCCS)	336/328	51.5± 5.9	26.8	33.3	50.9± 5.9	26.6	34.9	51.2± 5.1	26.8	34.3
Nurses' Health Study- Case-control study of type II diabetes (NHS-T2D)	-/1720	—	—	—	56.5 ± 6.9	27.1	53.5	—	—	—
Amish Family Osteoporosis Study (AFOS)	141/189	48.5± 13.9	25.8	54.3	49.4± 13.8	27.7	53.2	49.0 ± 13.9	26.8	53.8

*sample size based on available information on body mass index and 25(OH)D.

To replicate our findings on the association between the vitamin D-related SNPs and allele scores with BMI, we used the data from the genome-wide meta-analyses on BMI conducted as part of the Genetic Investigation of Anthropometric Traits (GIANT) consortium [23]. The GIANT meta-analyses consisted of 46 studies with up to 123,865 adults of European ancestry, including the 1958 British Birth Cohort, Framingham Heart study, Nurses' Health Study, Twins UK, UK Blood Services Common Control Collection, the Amish Family Osteoporosis Study, Health2000 GEN-METS sub-sample, and Northern Finland Birth Cohort 1966, which were also part of the D-CarDia collaboration.

7.2.3 GENOTYPING

We selected 12 established BMI-related SNPs (fat mass and obesity-associated, [FTO]- rs9939609, melanocortin 4 receptor [MC4R]- rs17782313, transmembrane protein 18 [TMEM18]- rs2867125, SH2B adaptor protein 1 [SH2B1]- rs7498665, brain-derived neurotrophic factor [BDNF]- rs4074134, potassium channel tetramerisation domain containing 15 [KCTD15]- rs29941, ets variant 5 [ETV5]- rs7647305, SEC16 homolog B [SEC16B]- rs10913469, Fas apoptotic inhibitory molecule 2 [FAIM2]- rs7138803, neuronal growth regulator 1 [NEGR1]- rs3101336, mitochondrial carrier 2 [MTCH2]- rs10838738, and glucosamine-6-phosphate deaminase 2 [GNPDA2]- rs10938397) for our analysis based on the study by Li et al. [24] and previously published genome-wide association studies for obesity-related traits [23],[25],[26]. The four vitamin D-related SNPs (DHCR7- rs12785878, CYP2R1- rs10741657, GC- rs2282679, and CYP24A1- rs6013897) were chosen on the basis of the recent genome-wide association study on 25(OH)D [27]. The studies that did not have genotyped data analysed imputed or proxy SNPs ($r^2 = 1$) as available (with a call threshold of 0.9 for the SNPs imputed with Impute; for those imputed with MACH, a call threshold of 0.8 was used) [28]. The genetic data for most studies were obtained from genome-wide association platforms, but for some studies, variants were genotyped de novo (MRC Ely, the Canadian Multicentre Osteoporosis Study, the Hertfordshire cohort study) or obtained through metabochip custom array (MRC Ely). Five studies did not have all the BMI-related SNPs (Framingham Heart Study [one missing SNP], Hertfordshire cohort study [three missing SNPs], InCHIANTI [two missing SNPs], PIVUS [two missing SNPs], and ULSAM [three missing SNPs]) and were still included in the BMI allele score analysis. Table S1 shows the minor allele frequencies for the BMI and vitamin D SNPs that were included in the analysis. A detailed description of the genotyping methods is provided in Text S2.

7.2.4 STATISTICAL ANALYSIS

Analyses in each study were performed according to a standardized analysis plan. When used as outcome variables, 25(OH)D and BMI were natural log transformed to be more closely approximated by normal distributions. If multiplied by 100, coefficients from linear regression models with ln transformed outcomes can be interpreted as the percentage difference in the outcome [29]. Models with BMI as an outcome were adjusted for age, sex, geographical site, and/or principal components from population stratification analysis (depending on data available); models with 25(OH)D as the outcome were additionally adjusted for month of blood sample collection (as a categorical variable) to account for seasonal variation and laboratory batch, where relevant. To assess the BMI relationship with 25(OH)D and vice versa, each study ran linear regression models adjusting for the covariates listed for each outcome, and the models were repeated stratifying by sex.

For the BMI SNPs, the effect allele was the BMI raising allele as established by Speliotes et al. [23]. We created a weighted score in each study [30], by multiplying each SNP (coded as 0–2) by a weight based on its effect size with BMI in the meta-analysis by Speliotes et al. [23]. The weighted BMI allele score was rescaled over the sum of weights for the available SNPs in each study to facilitate interpretation [30]. For the vitamin D SNPs, the effect allele was the 25(OH)D lowering allele as established by the SUNLIGHT Consortium [27]. As external weights were not available and the use of internal weights could bias the instrumental variable (IV) results [31], we performed an unweighted allele score analysis for the vitamin D SNPs. Vitamin D SNPs were used to form two separate allele scores [32]: a "synthesis" allele score, created by summing the risk alleles in *DHCR7* and *CYP2R1*, and a "metabolism" allele score, created by summing the risk alleles in *GC* and *CYP24A1* (Figure S1). Synthesis allele score was not created for the LURIC study (one missing SNP) and both synthesis and metabolism allele scores were not created for the MRC Ely study (two missing SNPs). The synthesis allele score included the SNPs that contribute directly to the production of 25(OH)D, and hence, for which the association with the outcome can be readily estimated based on the magnitude of the association between the score and 25(OH)D [32]. All analyses were done separately for the "metabolism" SNPs that are involved in the clearance or transport of 25(OH)D (with possible influences on bioavailability [33]) as the quantification of the association with the outcome based on the observed SNP-25(OH)D association is more difficult [32]. We also evaluated the joint contribution of synthesis and metabolism scores on BMI by including both vitamin D scores as separate variables in a multiple regression model. To examine the strength of the allele scores as instruments, the F-statistic was approximated from the proportion of variation in the respective phenotype (R^2) explained by the allele score, [F-stat = $(R^2 \times (n-2))/(1-R^2)$] [34].

To confirm our findings on the association between the vitamin D-related SNPs and allele scores with BMI in a larger sample, we used the summary statistics for the four vitamin D-related SNPs from the GIANT consortium. These SNPs were combined into synthesis and metabolism allele scores using an approximation method as previously described [35]. The individual SNP association with BMI is then weighted according to its predefined effect size and meta-analysed using the inverse-variance method with the other SNPs in the score [35]. The formal MR analyses to estimate the possible causal effect of BMI on 25(OH)D (and vice versa) were done using the IV ratio method [20],[36]. To estimate the IV ratio for the BMI effect on 25(OH)D, the meta-analysed association of the BMI allele score with 25(OH)D was divided by the association of BMI allele score with BMI. The variance for the IV ratio was estimated using a Taylor expansion [36]. The corresponding calculation was done to establish the 25(OH)D effect on BMI, with the IV ratio method applied separately for the two vitamin D allele scores. The joint contribution of the two vitamin D scores on BMI was assessed by multivariate meta-analysis [37], which incorporated the covariance matrix as estimated by study specific analyses.

In the presence of heterogeneity of association between the studies, random effects meta-analyses [38] were run, otherwise fixed effects models were used. Univariate meta-regression models were run to assess differences in the observed associations by study level factors of sex, average BMI (BMI≤25 kg/m² versus >25 kg/m²), the average age of participants (≤40, 41–60, and ≥61 y old), continent (North America versus Europe), and vitamin D assay (radio-immunoassay, enzyme-linked radio-immunoassay, and mass spectrometry). Power calculations for IV regression were performed by simulation [32] on the basis of associations observed between the phenotypes and their genetic proxies. For comparability across instruments/outcomes, power was determined for 0.02 log unit increase/decrease by decile, approximately corresponding to the association observed between BMI and 25(OH)D. To evaluate the ability to detect weaker effects on BMI using the synthesis and metabolism scores, power was also calculated for a 50% weaker effect (0.01 log unit increase/decrease). All meta-analyses and power calculations were performed at the Institute of Child Health (University College London, London) using STATA version 12 [39].

7.3 RESULTS

7.3.1 PHENOTYPIC ASSOCIATION BETWEEN BMI AND 25(OH)D CONCENTRATIONS

In the meta-analyses of 21 studies, each unit (kg/m²) increase in BMI was associated with 1.15% (95% CI 0.94%–1.36%, p = 6.52×10^{-27}) lower concentrations of 25(OH)D after adjusting for age, sex, laboratory batch, month of measurement, and principal components. The inverse association between BMI and 25(OH)D was stronger among the studies from North America than those from Europe (−1.58% [−1.81% to −1.36%], p = 1.01×10^{-43} versus −0.91% [−1.18% to −0.64%], p = 4.55×10^{-11}; pmeta-regression = 0.004) and for women than men (−1.43% [−1.65% to −1.22%],

$p = 1.13 \times 10^{-38}$ versus -0.75% $[-1.00\%$ to $-0.50\%]$, $p = 3.89 \times 10^{-9}$; $p_{\text{meta-regression}} = 4.10 \times 10^{-4}$) while no variation was seen by average age ($p_{\text{meta-regression}} = 0.78$) or BMI ($p_{\text{meta-regression}} = 0.48$) (Figure 1A and 1B).

7.3.2 EVALUATION OF CAUSAL ASSOCIATION USING MR APPROACH

The BMI allele score created from the 12 BMI-related SNPs showed a positive dose-response association with BMI (per unit increase 0.14% [0.12%–0.16%], $p = 6.30 \times 10^{-62}$), and both vitamin D allele scores showed the expected strong associations with 25(OH)D (per allele in synthesis score: -3.47% $[-3.90\%$ to $-3.05\%]$, $p = 8.07 \times 10^{-57}$; metabolism allele score: -5.38% $[-5.84\%$ to $-4.93\%]$, $p = 1.07 \times 10^{-118}$) (Figures 2, S2, and S3). The BMI allele score was also associated with 25(OH)D concentrations (per unit increase -0.06%, $[-0.10\%$ to $-0.02\%]$, $p = 0.004$) (Figure 3), while no association with BMI was seen for either the vitamin D synthesis or metabolism allele scores (per allele in synthesis score: 0.01% $[-0.17\%$ to 0.20%], $p = 0.88$, metabolism allele score: 0.17% $[-0.02\%$ to 0.35%], $p = 0.08$]) (Figure 4A and 4B). Analyses of joint effects by synthesis and metabolism scores provided no evidence for an association between 25(OH)D and BMI (per allele in synthesis score -0.03% $[-0.23\%$ to 0.16%] and metabolism score 0.17% $[-0.04\%$ to 0.37%], joint contribution $p = 0.26$).

In the analyses to establish the direction and causality of BMI–25(OH)D association by the use of the IV ratio, BMI was associated with 25(OH)D: each 10% increase in BMI lead to a 4.2% decrease in 25(OH)D concentrations (-7.1% to -1.3%; $p = 0.005$). However, the IV ratio analyses provided little evidence for a causal effect of 25(OH)D on BMI ($p \geq 0.08$ for both). We have summarised the coefficients for the MR analyses in Table 2.

The lack of association of the vitamin D allele scores with BMI was further confirmed using the GIANT consortium including 123,864 individuals in 46 studies [23]: neither the synthesis nor the metabolism allele score showed any evidence for an association with BMI ($p \geq 0.57$ for both) (Table 3).

A

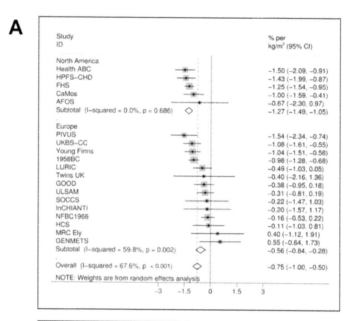

B

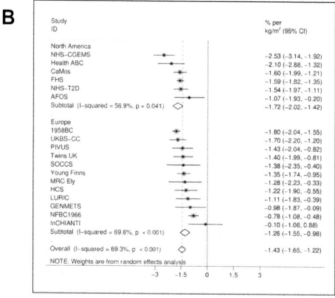

FIGURE 1: Random effects meta-analysis of the BMI association with 25(OH)D in men (A) (n = 20,950) and women (B) (n = 21,074). 95% confidence intervals given by error bars.

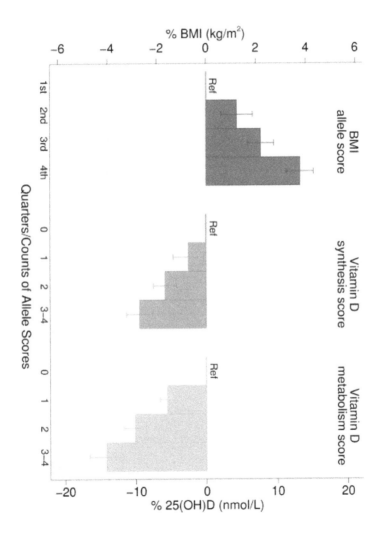

FIGURE 2: Meta-analysis of the BMI allele score association with BMI (n = 32,391), and the vitamin D synthesis (n = 35,873) and metabolism (n = 38,191) allele score association with 25(OH)D. 95% confidence intervals given by error bars.

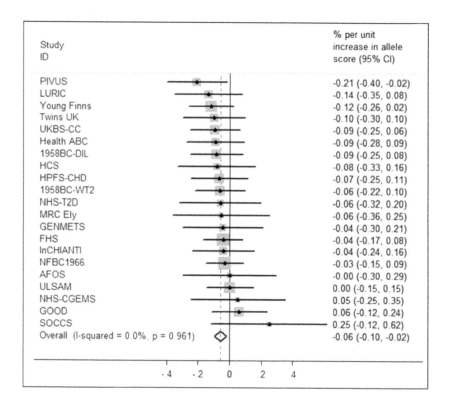

FIGURE 3: Meta-analysis of the BMI allele score association with 25(OH)D (n = 31,120). 95% confidence intervals given by error bars.

A

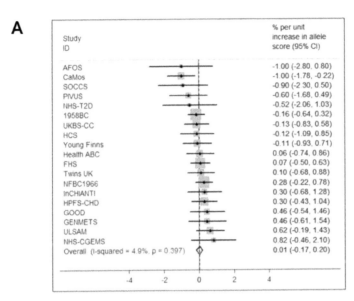

B

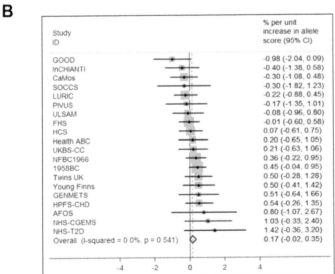

FIGURE 4: Meta-analysis of the synthesis allele score association with BMI (A) (n = 36,553) and the metabolism allele score association with BMI (B) (n = 40,367). 95% confidence intervals given by error bars.

TABLE 2: Summary of the coefficients used for IV ratio analyses.

IV	Allele Score with the Intermediate Trait	Allele Score with the Outcome	IV Ratio[a]	
	Coefficient, % (95% CI)	Coefficient,% (95% CI)	Coefficient (95% CI)	p-Value
BMI risk score	0.14 (0.12-0.16)	-0.06 (-0.1 to -0.02)	-0.42 (-0.71 to -0.13)	0.005
Synthesis score	-3.47 (-3.90 to -3.05)	0.01 (-0.17 to 0.20)	-0.00 (-0.06 to 0.05)	0.88
Metabolism score	-5.38 (-5.84 to -4.93)	0.17 (-0.02 to 0.35)	-0.03 (-0.06 to 0.01)	0.08

[a]*Calculated as the ratio between the allele score association with the outcome and intermediate trait. Coefficients can be interpreted as percent change in the outcome by percent change in the intermediate trait.*

7.3.3 ADDITIONAL ANALYSES

7.3.3.1 VALIDATION OF THE GENETIC INSTRUMENTS

The BMI SNPs and the vitamin D SNPs were all individually associated with BMI and 25(OH)D, respectively (Figures S4 and S5). The exception was KCTD15 SNP, which despite previous evidence for an association [25], was not associated with BMI in our meta-analyses. Across the studies, the 12 BMI SNPs combined as the BMI allele score explained 0.97% of the variation in BMI (F-statistic = 316; n = 32,391). The synthesis allele score explained 0.64% (F-statistic = 230; n = 35,873) and the metabolism allele score 1.26% (F statistic = 489; n = 38,191) of the variation in 25(OH)D. There was no evidence for variation in the BMI allele score–BMI association by continent ($p_{meta-regression}$ = 0.15) or BMI ($p_{meta-regression}$ = 0.83). However, the BMI allele score–BMI association was slightly weaker in studies with older compared to younger participants (−0.03% [−0.05% to −0.002%], $p_{meta-regression}$ = 0.03). The vitamin D allele score–25(OH)D association did not vary by age, BMI, continent, or assay ($p_{meta-regression} \geq 0.09$ for all comparisons).

TABLE 3: Results for the association between vitamin D SNPs/allele scores and BMI from the GIANT consortium.

SNPs/Allele Scores	Gene Symbol	Per Allele Change in BMI, kg/m^2	(95% CI) p-Value
rs12785878	*DHCR1*	0.001 (-0.01 to 0.009)	0.78
rs10741657	*CYP2R1*	-0.005 (-0.004 to 0.01)	0.30
Synthesis allele score (rs1278S878+rs10741657)	*DHCR1 + CYP2R1*	-0.002 (-0.009 to 0.005)[a]	0.57
rs2282679	*GC*	0.001 (-0.011 to 0.010)	0.91
rs6013897	*CYP24A1*	0.003 (-0.008 to 0.014)	0.61
Metabolism allele score (rs2282679+rs6013897)	*GC + CYP24A1*	0.002 (-0.006 to 0.009)[a]	0.67

The GIANT meta-analyses consisted of 46 studies with up to 123,86S adults of European ancestry [23], including the 1958 British Birth Cohort, Framingham Heart study, Nurses' Health Study, Twins UK, UK Blood Services Common Control Collection, the Amish Family Osteoporosis Study, Health2000 GENMETS sub-sample, and Northern Finland Birth Cohort 1966, which were also part of the D-CarDia collaboration. [a]Calculated as described in Ehret et al. [3S].

7.3.3.2 EVALUATION OF THE GENETIC OUTCOME ASSOCIATIONS

Of the 12 individual BMI SNPs, the SNP for FTO was the only one that showed evidence of a univariate association with 25(OH)D (p = 0.050) (Figure S6). None of the four 25(OH)D SNPs were individually associated with BMI (p≥0.10) (Figure S7). The lack of association of the four vitamin D SNPs with BMI was further confirmed using the summary data from the GIANT consortium (p>0.30 for all the SNPs) (Table 3).

The association between BMI allele score and 25(OH)D did not vary by study level factors, including age ($p_{meta-regression}$ = 0.40), BMI ($p_{meta-regression}$ = 0.18), continent of study ($p_{meta-regression}$ = 0.78), or vitamin D assay ($p_{meta-regression}$ = 0.23). Similarly, there was no evidence for variation in the vitamin D allele score–BMI association by age ($p_{meta-regression}$≥0.25 for both scores), or continent ($p_{meta-regression}$≥0.50 for both scores). There was also no strong evidence for variation in the vitamin D allele score–BMI association by average BMI of the study (≤ 25 kg/m^2 versus ≥ 25 kg/m^2), although for the synthesis score the meta-regression coefficient was of borderline significance ($p_{\text{meta-regression}}$ = 0.053, Figure S8; $p_{\text{meta-regression}}$ = 0.78 for metabolism score).

7.3.3.3 POWER COMPARISON

Illustrative power calculations are presented in Figure S9. In theory, we had greater power to detect an association between 25(OH)D and BMI using the metabolism score as an instrument, compared with an equal sized association between BMI and 25(OH)D using the BMI risk score. However, if the size of the association between 25(OH)D and BMI was only half that seen between BMI and 25(OH)D, our study would not have been adequately powered even with the inclusion of the GIANT results.

7.4 DISCUSSION

Obesity, and perhaps vitamin D deficiency, are among the most important modifiable risk factors for a number of chronic diseases. Obesity and vitamin D status are known to be associated but the direction of the association and whether it is causal has been uncertain. We have presented genetic evidence that higher BMI leads to lower vitamin D status. Conversely, our analyses provided no evidence for a causal role of vitamin D in the development of obesity, although our study was not powered to detect very small effects. These results suggest that although increases in vitamin D status are not likely to help with weight regulation, increased risk of vitamin D deficiency could contribute to the adverse health effects associated with obesity.

The association between obesity and vitamin D status was remarkably consistent across the different populations included in our meta-analyses, being apparent both in men and in women, and in the young and older cohorts alike. Interestingly, the association between obesity and 25(OH)D concentrations appeared stronger for populations in North America compared to Europe, possibly reflecting differences in the distribution of

BMI across the continents. Recent intervention studies have shown that obese individuals need higher vitamin D dosages than lean individuals to achieve the same 25(OH)D concentrations [40],[41]. Given that North America has one of the highest rates of obesity in the world [42], our study highlights the importance of considering obesity as a risk factor for vitamin D deficiency with implications on the dosage requirements and possible targeting of relevant health promotion strategies.

The lack of any suggestion for an association between the vitamin D SNPs and BMI in the GIANT consortium (n = 123,864) alongside our own large meta-analyses provides a strong case against linear increases in 25(OH)D having a substantive influence on BMI. This conclusion is in accordance with a recent study on Chinese women (n = 7,000), which also failed to observe evidence for an association with BMI for genetic variants in the vitamin D pathway [43]. Although a recent RCT (n = 77) suggested greater loss in fat mass for women receiving vitamin D [15], previous trials have failed to show any evidence for an effect despite larger treatment groups (n = 200–445), use of higher vitamin D dosages, and equal duration of treatment (12 mo) [13],[14]. Dilution related to the greater volume of distribution has been recently proposed as the most likely explanation for the lower 25(OH)D concentrations in obese individuals [44]. In that study, no evidence was found for reduced bioavailability through increased sequestration of vitamin D in the adipose tissue, which had previously been suggested to contribute to the low 25(OH)D concentrations in obesity [18]. In contrast, intact parathyroid hormone (iPTH) levels [45], which stimulate the 1-α-hydroxylase (CYP27B1) enzyme that converts 25(OH)D to 1,25-dihydroxyvitamin D (the active hormonal form), have been found to be elevated in obesity [46], which could to some extent also contribute to the lower 25(OH)D concentrations in obese individuals. It is also possible that differences in lifestyle could contribute to lower 25(OH)D concentrations in obese compared to normal weight individuals, although the association between obesity and low 25(OH)D concentrations has been found to only modestly attenuate after adjustment for vitamin D-related lifestyle and dietary factors [9].

The main strengths of this study are the large sample size and the individual level population-based data from North America and Europe. We used a bi-directional MR approach to investigate the causal directions between obesity and vitamin D deficiency, observing evidence for reductions in 25(OH)D by BMI but not vice versa. However, based on the biological pathways proposed, a possible effect of 25(OH)D on BMI could be expected to be weaker than the effect of BMI on 25(OH)D. Despite including data from the large GIANT consortium to narrow the range of effects compatible with the data, we are unable to exclude very small effects. Furthermore, while the MR approach enables the approximation of life-long differences in average concentrations, with genetic markers it is not possible to examine the influences arising from the extremes of non-linear distributions [20]. Consequently, we cannot discount a possible effect of severe vitamin D deficiency on BMI due to evidence of non-linearity seen in some studies [47]. In contrast, associations between BMI and 25(OH) D within levels in the obesity range were consistently linear in studies included in our analyses (unpublished data), hence the observed association between higher BMI and lower 25(OH)D is likely to be informative in the context of obesity.

One of the methodological challenges of the MR approach relates to the large sample size requirement, arising from the availability of relatively weak instruments for most exposures [22],[31]. This aspect of the MR approach is also reflected in our study, notably in the relatively small amount of variation explained by all the instruments used. We used the IV ratio method on meta-analyzed coefficients since all studies were not able to share individual level participant data. This method assumes linear relationships and may have less power to detect an effect than other IV methods [48]. However, as shown by the clear outcome of these analyses, we were able to overcome these issues by combining several cohorts with comparable information, allowing us to achieve the large numbers required (maximum n = 42,024) [31]. To confirm the lack of association between vitamin D-related genetic variations and BMI, we were able to expand the analyses by using data from the large GIANT meta-analyses (n = 123,864) [23]. However, this cannot be considered an independent replication, as eight of the studies that were part of the D-CarDia Collaboration were also included in GIANT. The F-statistic is used to measure the strength of an instrument, and an instrument that has a value greater than 10 is considered strong enough to use in IV analyses [49]. In our analyses, the F-statistic was greater than 200 for all instruments used due to our large sample size.

Combining large population-based studies from North America and Europe could lead to confounding by population stratification; however, we adjusted for geographical variation/principal components in all analyses, which appeared adequate, as there was no evidence for heterogeneity by continent for the allele score meta-analyses. An important benefit of the MR approach is that it helps to overcome problems of confounding and reverse causality, which limit the ability to draw causal inferences in non-genetic observational studies [19],[20]. However, it could be argued that as the biological function for some of the BMI SNPs is yet to be established, there could be alternative biological pathways explaining their association with BMI. Using multiple SNPs to index BMI, we were able to minimise the risk of pleiotropic effects, as the effects of alternative pathways reflected by individual SNPs would be expected to be strongly diluted when combined in a multi marker score [21],[22].

In conclusion, we demonstrated that the association between BMI and lower 25(OH)D concentrations in Caucasian populations from North America and Europe can be seen across different age groups and in both men and women. We also show that higher BMI leads to lower vitamin D status, providing evidence for the role of obesity as a causal risk factor for the development of vitamin D deficiency. Together with the suggested increases in vitamin D requirements in obese individuals [45],[50], our study highlights the importance of monitoring and treating vitamin D deficiency as a means of alleviating the adverse influences of excess adiposity on health. Our findings suggest that population level interventions to reduce obesity would be expected to lead to a reduction in the prevalence of vitamin D deficiency.

REFERENCES

1. 1. Baskin ML, Ard J, Franklin F, Allison DB (2005) Prevalence of obesity in the United States. Obes Rev 6: 5–7. doi: 10.1111/j.1467-789x.2005.00165.x
2. Ogden CL, Carroll MD, Curtin LR, Lamb MM, Flegal KM (2010) Prevalence of high body mass index in US children and adolescents, 2007–2008. JAMA 303: 242–249. doi: 10.1001/jama.2009.2012

3. Berghofer A, Pischon T, Reinhold T, Apovian CM, Sharma AM, et al. (2008) Obesity prevalence from a European perspective: a systematic review. BMC Public Health 8: 200. doi: 10.1186/1471-2458-8-200

4. Zheng W, McLerran DF, Rolland B, Zhang X, Inoue M, et al. (2011) Association between body-mass index and risk of death in more than 1 million Asians. N Engl J Med 364: 719–729. doi: 10.1056/nejmoa1010679

5. Flegal KM, Carroll MD, Kit BK, Ogden CL (2012) Prevalence of obesity and trends in the distribution of body mass index among US adults, 1999–2010. JAMA 307: 491–497. doi: 10.1001/jama.2012.39

6. Vimaleswaran KS, Loos RJ (2010) Progress in the genetics of common obesity and type 2 diabetes. Expert Rev Mol Med 12: e7. doi: 10.1017/s1462399410001389

7. Ginde AA, Liu MC, Camargo CA Jr (2009) Demographic differences and trends of vitamin D insufficiency in the US population, 1988–2004. Arch Intern Med 169: 626–632. doi: 10.1001/archinternmed.2008.604

8. Lanham-New SA, Buttriss JL, Miles LM, Ashwell M, Berry JL, et al. (2011) Proceedings of the Rank Forum on vitamin D. Br J Nutr 105: 144–156. doi: 10.1017/s0007114510002576

9. Hyppönen E, Power C (2007) Hypovitaminosis D in British adults at age 45 y: nationwide cohort study of dietary and lifestyle predictors. Am J Clin Nutr 85: 860–868.

10. Earthman CP, Beckman LM, Masodkar K, Sibley SD (2012) The link between obesity and low circulating 25-hydroxyvitamin D concentrations: considerations and implications. Int J Obes (Lond) 36: 387–396. doi: 10.1038/ijo.2011.119

11. Shi H, Norman AW, Okamura WH, Sen A, Zemel MB (2001) 1alpha,25-Dihydroxyvitamin D3 modulates human adipocyte metabolism via nongenomic action. Faseb J 15: 2751–2753. doi: 10.1096/fj.01-0584fje

12. Fassina G, Maragno I, Dorigo P, Contessa AR (1969) Effect of vitamin D2 on hormone-stimulated lipolysis in vitro. Eur J Pharmacol 5: 286–290. doi: 10.1016/0014-2999(69)90150-2

13. Sneve M, Figenschau Y, Jorde R (2008) Supplementation with cholecalciferol does not result in weight reduction in overweight and obese subjects. Eur J Endocrinol 159: 675–684. doi: 10.1530/eje-08-0339

14. Zittermann A, Frisch S, Berthold HK, Gotting C, Kuhn J, et al. (2009) Vitamin D supplementation enhances the beneficial effects of weight loss on cardiovascular disease risk markers. Am J Clin Nutr 89: 1321–1327. doi: 10.3945/ajcn.2008.27004

15. Salehpour A, Shidfar F, Hosseinpanah F, Vafa M, Razaghi M, et al. (2012) Vitamin D3 and the risk of CVD in overweight and obese women: a randomised controlled trial. Br J Nutr 1–8. doi: 10.1017/s0007114512000098

16. Soares MJ, Murhadi LL, Kurpad AV, Chan She Ping-Delfos WL, Piers LS (2012) Mechanistic roles for calcium and vitamin D in the regulation of body weight. Obes Rev 13: 592–605. doi: 10.1111/j.1467-789x.2012.00986.x

17. Foss YJ (2009) Vitamin D deficiency is the cause of common obesity. Med Hypotheses 72: 314–321. doi: 10.1016/j.mehy.2008.10.005

18. Wortsman J, Matsuoka LY, Chen TC, Lu Z, Holick MF (2000) Decreased bioavailability of vitamin D in obesity. Am J Clin Nutr 72: 690–693.

19. Davey Smith G, Ebrahim S (2003) 'Mendelian randomization': can genetic epidemiology contribute to understanding environmental determinants of disease? Int J Epidemiol 32: 1–22. doi: 10.1093/ije/dyg070

20. Lawlor DA, Harbord RM, Sterne JA, Timpson N, Davey Smith G (2008) Mendelian randomization: using genes as instruments for making causal inferences in epidemiology. Stat Med 27: 1133–1163. doi: 10.1002/sim.3034

21. Davey Smith G (2011) Random allocation in observational data: how small but robust effects could facilitate hypothesis-free causal inference. Epidemiology 22: 460–463; discussion 467–468. doi: 10.1097/ede.0b013e31821d0426

22. Palmer TM, Lawlor DA, Harbord RM, Sheehan NA, Tobias JH, et al. (2012) Using multiple genetic variants as instrumental variables for modifiable risk factors. Stat Methods Med Res 21: 223–242. doi: 10.1177/0962280210394459

23. Speliotes EK, Willer CJ, Berndt SI, Monda KL, Thorleifsson G, et al. (2010) Association analyses of 249,796 individuals reveal 18 new loci associated with body mass index. Nat Genet 42: 937–948.

24. Li S, Zhao JH, Luan J, Luben RN, Rodwell SA, et al. (2010) Cumulative effects and predictive value of common obesity-susceptibility variants identified by genome-wide association studies. Am J Clin Nutr 91: 184–190. doi: 10.3945/ajcn.2009.28403

25. Loos RJ, Lindgren CM, Li S, Wheeler E, Zhao JH, et al. (2008) Common variants near MC4R are associated with fat mass, weight and risk of obesity. Nat Genet 40: 768–775.

26. Thorleifsson G, Walters GB, Gudbjartsson DF, Steinthorsdottir V, Sulem P, et al. (2009) Genome-wide association yields new sequence variants at seven loci that associate with measures of obesity. Nat Genet 41: 18–24. doi: 10.1038/ng.274

27. Wang TJ, Zhang F, Richards JB, Kestenbaum B, van Meurs JB, et al. (2010) Common genetic determinants of vitamin D insufficiency: a genome-wide association study. Lancet 376: 180–188. doi: 10.1016/s0140-6736(10)60588-0

28. Zheng J, Li Y, Abecasis GR, Scheet P (2011) A comparison of approaches to account for uncertainty in analysis of imputed genotypes. Genet Epidemiol 35: 102–110. doi: 10.1002/gepi.20552

29. Cole TJ (2000) Sympercents: symmetric percentage differences on the 100 log(e) scale simplify the presentation of log transformed data. Stat Med 19: 3109–3125. doi: 10.1002/1097-0258(20001130)19:22<3109::aid-sim558>3.0.co;2-f

30. Lin X, Song K, Lim N, Yuan X, Johnson T, et al. (2009) Risk prediction of prevalent diabetes in a Swiss population using a weighted genetic score–the CoLaus Study. Diabetologia 52: 600–608. doi: 10.1007/s00125-008-1254-y

31. Pierce BL, Ahsan H, Vanderweele TJ (2011) Power and instrument strength requirements for Mendelian randomization studies using multiple genetic variants. Int J Epidemiol 40: 740–752. doi: 10.1093/ije/dyq151

32. Berry DJ, Vimaleswaran KS, Whittaker JC, Hingorani AD, Hypponen E (2012) Evaluation of genetic markers as instruments for mendelian randomization studies on vitamin D. PLoS One 7: e37465 doi:10.1371/journal.pone.0037465.

33. Chun RF, Lauridsen AL, Suon L, Zella LA, Pike JW, et al. (2010) Vitamin D-binding protein directs monocyte responses to 25-hydroxy- and 1,25-dihydroxyvitamin D. J Clin Endocrinol Metab 95: 3368–3376. doi: 10.1210/jc.2010-0195

34. Rice JA (1995) Expected values. Mathematical statistics and data analysis. 2nd edition. Pacific Grove (California): Duxbury Press.
35. Ehret GB, Munroe PB, Rice KM, Bochud M, Johnson AD, et al. (2011) Genetic variants in novel pathways influence blood pressure and cardiovascular disease risk. Nature 478: 103–109.
36. Thomas DC, Lawlor DA, Thompson JR (2007) Re: Estimation of bias in nongenetic observational studies using "Mendelian triangulation" by Bautista et al. Ann Epidemiol 17: 511–513. doi: 10.1016/j.annepidem.2006.12.005
37. White IR (2009) Multivariate random-effects meta-analysis. The Stata Journal 9: 40–56.
38. Borenstein M (2009) Introduction to meta-analysis. Chichester: John Wiley & Sons. xxviii.
39. StataCorp (2011). Stata Statistical Software: Release 12: College Station (Texas): StataCorp LP.
40. Jorde R, Sneve M, Emaus N, Figenschau Y, Grimnes G (2010) Cross-sectional and longitudinal relation between serum 25-hydroxyvitamin D and body mass index: the Tromso study. Eur J Nutr 49: 401–407. doi: 10.1007/s00394-010-0098-7
41. Lee P, Greenfield JR, Seibel MJ, Eisman JA (2009) Center JR (2009) Adequacy of vitamin D replacement in severe deficiency is dependent on body mass index. Am J Med 122: 1056–1060. doi: 10.1016/j.amjmed.2009.06.008
42. Bassett DR Jr, Pucher J, Buehler R, Thompson DL, Crouter SE (2008) Walking, cycling, and obesity rates in Europe, North America, and Australia. J Phys Act Health 5: 795–814.
43. Dorjgochoo T, Shi J, Gao YT, Long J, Delahanty R, et al. (2012) Genetic variants in vitamin D metabolism-related genes and body mass index: analysis of genome-wide scan data of approximately 7000 Chinese women. Int J Obes (Lond) 36: 1252–1255. doi: 10.1038/ijo.2011.246
44. Drincic AT, Armas LA, Van Diest EE, Heaney RP (2012) Volumetric dilution, rather than sequestration best explains the low vitamin D status of obesity. Obesity (Silver Spring) 20: 1444–1448. doi: 10.1038/oby.2011.404
45. Bell NH, Epstein S, Greene A, Shary J, Oexmann MJ, et al. (1985) Evidence for alteration of the vitamin D-endocrine system in obese subjects. J Clin Invest 76: 370–373. doi: 10.1172/jci111971
46. Holick MF (2007) Vitamin D deficiency. N Engl J Med 357: 266–281. doi: 10.1056/nejmra070553
47. Hyppönen E, Berry D, Cortina-Borja M, Power C (2010) 25-Hydroxyvitamin D and pre-clinical alterations in inflammatory and hemostatic markers: a cross sectional analysis in the 1958 British Birth Cohort. PLoS One 5: e10801 doi:10.1371/journal.pone.0010801.
48. Burgess S, Thompson SG, Andrews G, Samani NJ, Hall A, et al. (2010) Bayesian methods for meta-analysis of causal relationships estimated using genetic instrumental variables. Stat Med 29: 1298–1311. doi: 10.1002/sim.3843
49. Staiger D, Stock JH (1997) Instrumental variables regression with weak instruments. Econometrica 65: 557–586. doi: 10.2307/2171753

50. Huh SY, Gordon CM (2008) Vitamin D deficiency in children and adolescents: epidemiology, impact and treatment. Rev Endocr Metab Disord 9: 161–170. doi: 10.1007/s11154-007-9072-y

There are several supplemental files that are not available in this version of the article. To view this additional information, please use the citation on the first page of this chapter.

CHAPTER 8

The Relationship between Folic Acid and Risk of Autism Spectrum Disorders

YASMIN NEGGERS

8.1 INTRODUCTION

Autism, also referred to as autistic spectrum disorder (ASD) and pervasive developmental disorder (PDD), defines a group of neurodevelopmental disorders affecting approximately 1% of the population which is usually diagnosed in early childhood [1]. Since there are no definitive biological markers of autism for a majority of cases, diagnosis depends on a range of behavioral signs. Experts disagree about the causes and significance of the recent increases in prevalence of ASD [2]. Despite hundreds of studies, it is still not known why autism incidence increased rapidly during the 1990s and is still increasing in the 2000's [3]. The findings from updated (March 2014) population-based estimates from the Autism and Developmental Monitoring Network Surveillance (ADDM) in multiple U.S. communities, as reported by the Centers of Disease Control and Prevention (CDC), indicates an overall ASD prevalence of 14.7 per 1000 (95% C.I. = 14.3–5.1) or one in 68 children aged 8 years during 2010 [4]. This latest prevalence estimate of ASD as one in 68 children aged 8 years, was 29%

higher than the preceding estimate of one in 88 children or 11.3 per 1000 (95% C.I. = 11.0–11.7) [4].

Both genetic and environmental research has resulted in recognition of the etiologic complexity of ASD. An integrated metabolic profile that reflects the interaction of genetic, epigenetic, environmental and endogenous factors that disturb the pathway of interest needs to be evaluated [5]. Though it is established that ASD is a multi-factorial condition involving both genetic and a wide range of environmental risk factors, only during the past decade has the research into environmental risk factors grown significantly [6]. The contribution from environmental factors was originally thought to be low partly due to high monozygotic twin concordance in earlier studies and partly due to a limited understanding of gene-environment interactions [7]. Over the past 10 years, studies with biological plausible pathways, focused on critical time periods of neurodevelopment have resulted in promising risk and protective factors. One such area of research concerns potentially modifiable nutritional risk factors. Despite a number of studies evaluating the diet and nutritional status in children affected with ASD, there is still a paucity of research directly investigating the association between maternal nutrition and risk of ASD in the offspring [6,7]. Maternal nutrition is essential to fetal brain development, and maternal nutrient deficiencies have been associated with significant increased risk of various adverse neurodevelopmental outcomes, including neural tube defects and schizophrenia [8]. Fetal brain development in terms of structure and function has been shown to be influenced by maternal nutrient balance and deprivation, particularly common during pregnancy due to increased metabolic demands of the growing fetus, as well as increased nutrient needs of maternal tissues [9,10]. Therefore, it is quite plausibile that maternal nutritional status before and during pregnancy may influence ASD risk.

In this article the association between maternal folic acid intake, including folic acid supplementaion and risk of ASD in the offspring will be evaluated. It is of interest that though there is some evidence that folate intake during pregnancy decreases the risk of ASD [8,9,10], several investigators have speculated that a high maternal folate intake due to folic acid fortification of foods may be linked to increased prevalence of ASD [11,12,13,14].

8.2 FOLATE METABOLISM

Folate is a generic term for a vitamin, which includes naturally occuring food folate (reduced form, largely polyglutamated 5-methyltetrahydrofolate) and folic acid (oxidized form, pteroyl-L-monoglutamic acid) in supplements and fortified foods [15]. Folate has many coenzyme roles that function in the acceptance and transfer of 1-C units. The function of folate in mammals is to aquire single-carbon units, usually from serine, and transfer them in purine and pyrimidine biosynthsis; hence folate coenzymes are essential for synthesis of DNA [11,15]. Folate coenzymes are also necessary for de novo methionone synthesis and several other cellular components. Figure 1 illustrates how the folate cycle facilitates nucleic acid synthesis and is responsible for transfer of 1-C methyl groups to DNA and proteins [16]. Methyl groups added onto cytosine residues in the promoter region CpGs in genomic DNA are central to regulation of gene expression [17,18]. Recent investigations have led to suggestion that children with autism may have altered folate or methionine metabolism resulting in hypotheses that the folate-methionine cycle may play a key role in the etiology of autism. Main et al. conducted a systematic review to examine the evidence for the involvement of alterations in folate methionine metabolism in the etiology of autism [18]. The findings of the review of studies reporting data for metabolites, interventions or genes of the folate-methionine pathway and their related polymorphism were conflicting [18]. There was suggestion that changes in concentrations of metabolites of the methionine cycle may be driven by abnormalities in folate transport and/or metabolism. Most genetic studies lacked sufficient power to provide conclusive genetic relations. These investigators concluded that further research is needed before any definitive conclusions can be made about the role for a dysfunctional pathway in the etiology of autism.

Though an evaluation of the entire metabolic pathway of folic acid metabolism along with various genetic varitions in enzymes involved in the numerous pathways will provide greater mechanistic insights in the ASD pathology, it is beyond the scope of this article to do so. A targeted approach focusing on the role of folic acid in DNA methylation and recent epigenitic evidence will be discussed.

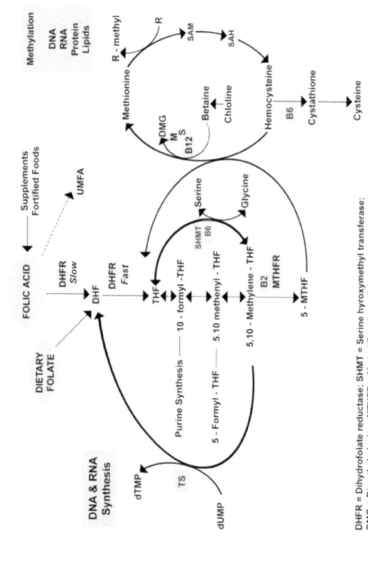

FIGURE 1: Folate metabolism [16].

DHFR = Dihydrofolate reductase; SHMT = Serine hyroxymethyl transferase; GMG = Dimethyl glycine; MTHFR = Methyl Tetrahydrofolate reductase. Adapted from[16]

8.3 DNA METHYLATION AND THE EPIGENETIC ROLE OF FOLIC ACID IN ASD

Coenzyme 5-methy tetrahydrofolate polyglutamate (5-CH3THFR) is needed by the salvage pathway to convert homocystiene to methionine [15]. Methionine can be converted to S-adenosyl methionine (SAM), a molecule with many roles, including methylation of cytosine residue in DNA and of arginine and lysine residues in histones, both of which are involved in regulating gene expression [13]. In the mammalian genome, methylation only occurs on cytocine residues that occur 5' to a guanosine residue in a CpG dinucleotide. CpG dinucleotides are enridched in "CpG islands" which are found proximal to promoter regions of about half the genes in the genome, and are primarily unmethylated. Methylation of promoter related Cpg islands can supress gene expression by causing chromatin condensation. DNA methylation is important as an epigenetic determinant of gene expression, in the maintainence of DNA integrity and in development of mutations [19]. Errors in normal epigenetic processes are called epimutations, and result in epigenetic silencing of a gene that is not normally silent. Low folate status is often associated with impairment of DNA methylation, but sometimes it leads to hypermethylation and could affect gene expression in complex ways (-19). From these studies, it is not clear whether an excess of folate might have adverse effects on these mutations.

There is speculation that folic acid supplementation may be associated with some aberrant conditions in children [18]. Junaid et al. reported that exposure of lymphoblastoid cells to folic acid supplementation causes widespread changes in gene expression [13]. Furthermore, Barua et al. [16] suggested that the occurrence of such epigenetic changes during gestational development may impact methylation status of DNA in the offspring's brain and cause altered gene expression, and since gestational development involves a highly orchestrated regulation of gene expression, this gene dysregulation may affect brain development and may result in various neuropsychiatric conditons. Baura et al. [17] identified substantial differently methylated regions (DMRs) in the the cerebral hemispheres of mice offspring of mothers on high folic acid intake as compared to offspring of mice on low maternal folic acid diet. These results support the findings of numerous studies that show that abnormalities in the frontal lobes impact brain development and autism [20,21,22,23,24]. Until very recently, changing the folate status in humans has been shown to influence DNA methylation, but it was not established whether alteration in DNA methylation after changes in folate status are harmful in humans. However, in a challenging study, Wong et al., performed a genome-wide analysis of DNA methylation in a sample of 50 monozygotic (MZ) twin pairs sampled from a representatitive population cohort that included twins discordant and concordant for ASD, ASD-associated traits and no autistic phenotype [20]. Within-twin and between group analyses identified numerous differently methylated regions assocated with ASD. These researchers also indicated significant correlations between DNA methylation and quantatively measured autistic trait scores for the sample cohort. This is the first systematic epigenomic analysis of MZ twin discordant for ASD and implicates a role for the altered DNA methylation in autism. Findings by Wong and colleagues provide support for a potential role of DNA methylation via a high maternal folic acid intake in ASD and ASD-related traits [20].

8.4 IS FOLIC ACID FORTIFICATION LINKED WITH INCREASED PREVALENCE OF ASD?

Beginning 1 January 1998, FDA mandated fortification of manufactured cereal products went into effect and since then there has been a significant decrease in cases of neural defects in the U.S. [25,26]. Several researchers have questioned whether an increase in maternal folate early during pregnancy might be partly related to the unexplained increase in ASD cases in the U.S. It is interesting that closure of the neural tube and therefore its enablement by folic acid supplementation, occurs at a time during embryogenesis that is also critical in autism development [11,27,28,29,30]. It is well established that there has been a significant enhancement of maternal folate status since FDA mandated folic acid fortification of certain foods (1998) which has resulted in decreased incidence of NTDs during the mid-2000s [15]. This same time period coincides with the apparent beginning and continuous rise in the prevalence of autism and related disorders in the U.S. Investigators have wondered whether these similar time frames of change in maternal folate status and possible autism prevalence are a random event or that the improved maternal and resulting fetal folate status has played a role [11]. The enzyme dihydrofolate reductase (DHFR) is necessary to convert dietary folic acid to tetrahydrofolate before its one-carbon derivatives can be used in the body as coenzymes for various methylation reactions and nucleotide synthesis [15]. Bailey and Ayling have reported that the process of reduction of folic acid to tetrahydrofolate in humans is slow and highly variable [31]. They showed that in human liver, reduction of folic acid by DHFR on average was less than 2% of that of rat liver. Also, folic acid is an inhibitor of DHFR in reduction of its substrate 7,8 dihydrofolic acid. This limited ability to activate the synthetic vitamer raises questions about use of high levels of folic acid. Concern has been expressed that this unmetabolized folic acid may be detrimental [11]. Thus, it can be speculated that some mothers of children with ASD may have unusually low activity of DHFR.

Rogers and other researchers [11,28,29,30] have explored the possibility that a particular polymorphic form of the key enzyme methylenetetrahydratefolate reductase (MTHFR), required for activation of folate for methylation in neurodevelopment, exhibits reduced activity under low or normal folate levels but normal activity under higher folate nutritional status. In several studies, higher plasma homocysteine levels than in non-carriers, resulting from the presence of polymorphic forms of MTHFR during reduced or normal folate status have been shown to result in increased rates of miscarriages via thrombotic effects [28,30]. However, under the condition of enhanced folate status during the perinatal period, the incidence of hyper-homocysteinemia is reduced and thereby masks the latent adverse effects of the presence of this polymorphic form of MTHFR during pregnancy [32]. This polymorphism, although common in the normal population, is found with significantly higher frequency in children with autism [30]. Thus, it is hypothesized that enhanced folate status during pregnancy from fortification could have increased the survival rate of fetuses with genetic polymorphism such as MTHFR 667 C > T, which are associated with high homocysteine and subsequently require higher amounts of folate for the normal methylation needed for proper neurodevelopment [32]. Such polymorphisms have been observed in higher frequencies in children with autism, suggesting that these children might be genetically predisposed to less efficient folate metabolism and function [30,32]. Haggerty et al. [29] conducted a study to evaluate the concern that increasing folic acid intake through fortification may select for embryos with genotype that increase the risk of disease like autism in the offspring. They found no evidence to support that folic acid fortification or supplement use in pregnancy results in selection of deleterious genotype.

8.5 DECREASED RISK OF ASD WITH IMPROVED MATERNAL FOLATE STATUS

Since folate and folic acid are essential for basic cellular processes, including DNA replication and protein methylation, it is biologically plausible that folic acid intake might affect numerous conditions positively or negatively depending on timing and dose. Several investigators have put forward hypotheses to explain the mechanism of association between folic acid and autism [33]. Ramaekers et al. [34,35] identified reduced 5-methylenetetrahydrofolate (5-Methyl THF) transport into the cerebrospinal fluid (CSF) in two autism spectrum disorders, i.e., Rett syndrome and infantile low-functioning autism due to folate receptor autoimmunity. In spite of normal serum folate, CSF 5-Methyl THF was low in 23 of 25 patients and was explained by serum folate receptor autoantibodies (FRA) blocking the folate binding site of the membrane attached FR on the choroid epithelial cell. A partial or complete clinical recovery was reported after 12 months of oral folinic acid supplements in affected children. These researchers suggested that FR autoimmunity and cerebral folate deficiency appear to play a crucial role in the pathogenesis of autism spectrum disorders or in a particular subgroup of the autism spectrum [35]. In an open label study conducted by Frye et al., 93 children with autism also had a high prevalence (75.3%) of folate receptor antibodies (FRA) [36], In 16 children, the concentration of FRAs significantly correlated with cerebrospinal fluid 5-MTHFA, which were below the normative mean in every case. Children with FRAs were treated with oral leucovorin calcium (50 mg/day), and treatment response compared with the wait list control group. Compared to controls, significantly higher improvement ratings were observed in treated children over a 4 month period in verbal communications, receptive and expressive language and stereotypical behavior. This study further supports the role of folic acid as a risk factor for autism. Many recent studies have pointed to improved neurodevelopment in autistic children with mothers having higher folate concentrations or receiving folic acid supplements [10]. These studies conducted in America, Europe, Asia, and South Asia have shown consistent positive effects of maternal folate status in reducing the risk of autism in the offspring [9,37,38,39,40,41,42]. In an extensive study Adams et al. [37] compared the nutritional (vitamins, minerals and amino acids) and metabolic status (biomarkers of oxidative stress, methylation and sulfuation) of children with autism with that of healthy neurotypical children to evaluate the association of autism severity with nutrient biomarkers. Though plasma folic acid concentrations were not significantly different in autistic cases as compared to those of controls, a biomarker of functional need for folic acid, average FIGLU concentration, was significantly higher in children with autism as compared to neurotypical controls (1.99 µg/L ± 0.92 vs. 1.62 µg/L ± 072). Also, S-adenosyl methionine (SAM), the primary methyl donor in the body, was also significantly lower in children with autism ($p < 0.001$). James et al., conducted studies indicating higher vulnerability to oxidative stress and a decreased capacity for methylation which may contribute to development of autism [5,41]. In two studies, James et al., also reported significant improvement in transmethylation metabolites and glutathione redox status in autistic children after treatment for 3 months with oral supplements of 800 µg folinic acid and 1000 mg betaine twice/day and 400 µg of folinic acid twice/day and 75 µg/kg methyl B_{12} twice a week respectively [5,41].

Following the positive results associated with maternal folate status and reduced risk of ASD in various case-control studies, Suren et al., conducted an excellent epidemiologically sound study which confirmed the association between maternal use of prenatal folic acid supplements and subsequent decreased risk of ASD in children [8]. Suren et al., evaluated 85,176 children from the Norwegian Mother and Child cohort Study (MoBa) and reported an incidence of ASD of 0.10% in offspring of mothers who took periceptional folic acid supplements as compared to 0.21% in offspring of those who did not. No foods were fortified with folic acid at the time of recruitment of subjects for this study; therefore synthetic supplements represented the only source of folate other than that from the diet for the pregnant women. The subjects for this study kept a record of intake of vitamins, minerals, and other supplements as listed on the ingredient lists on the supplement containers within 4 week intervals from the start of pregnancy. In addition, to quantify supplement use and dietary intake, a food frequency questionnaire was administered at mid-pregnancy. This study has several strengths such as prospective design, use of validated instruments to collect data on supplement use during pregnancy, and active screening of children for autism and other neurodevelopmental disorders [8]. A limited evidence of selection bias was indicated by comparing their results in MoBa with Norwegians Medical registry data on risk of autistic disorder in folic acid supplement users vs. non-users and finding similar results. Furthermore, the prevalence of ASD was lower in MoBa than in the U.S. but similar to the prevalence in Norway. Beaudet [43] in response to discovery of inborn errors of metabolism associated with autism discussed the possibility of preventable forms of autism. Folic acid supplementation might correct or ameliorate underlying genetic variation in children, their parents, or both might drive the observed reduction in risk reported by Surin et al. [8]. Such genetic variation could include alterations in epigenetic regulator genes and their targets, which have been previously associated with risk of ASD.

Several epidemiological studies have tested the above mentioned hypotheses regarding the association between folic acid/folate and ASD. In this section studies are reported in chronological order and by study design. Relatively recent studies (starting in 2000) with well-designed methodology are included to examine the evidence for the involvement of folic acid/folate or alteration in folate metabolism and risk of ASD. A summary of these studies is presented in Table 1 [44]. With the exception of a prospective study in a Norwegian children cohort [8] and an open label trial [41] most of these investigations consist of case control designs, where maternal perinatal folate or multivitamin supplementation of children with ASD was retrospectively compared with maternal perinatal folate or multivitamins supplement use by healthy children.. A few investigators have also evaluated the effects of folate metabolites and their possible role as oxidative stressors as a risk factor for autism and the effect of interaction between maternal folate status and maternal genotype and risk of autism

[12,28,39]. Results of these studies indicate an association between maternal perinatal folate status and ASD. Significant interaction effects have been reported for maternal MTHFR 677 TT, CBSrs234715 GT +TT, and child COMT 472 AA genotype, with greater risk for autism when mothers did not report taking prenatal vitamins peri-conceptionally. Schmidt et al. [9] have observed greater risk for children whose mothers had other one-carbon metabolism pathway gene variants and no maternal vitamin intake. As stated earlier, most of these investigations were retrospective case control studies with possibilities of various types of biases including differential misclassification of disease or/and exposure.

Currently several randomized clinical trials are underway to clarify and confirm the association between periconceptional folic acid intake and autism [45,46]. The Chinese Children and Family Study will evaluate potential benefits and adverse effects of periconceptional folic acid supplements in a 15 year follow-up of offsprings and mother [46]. In another open label clinical trial, with 40 autistic children, efficacy of methylcobalamine and folinic acid supplements would be determined. Whether treatment with these metabolic precursors would improve plasma bio-makers of oxidative stress and measures of core behaviors will be evaluated [45].

8.6 CONCLUSIONS

Whether perinatal folic acid supplementation can prevent autism is still an open question. Results of several recent studies, including the prospective study by Surin et al. [8] are encouraging, but it is too early to say that universal periconceptional use of folic can reduce the incidence of ASD resulting from abnormal folate- methionine metabolism. Some very recent epigenetic studies in monozygotic twins provide support for potential role of DNA methylation via a high maternal folic acid intake in ASD [19,20]. Specifically, findings by Wong and colleagues [20] are provocative and provide support for a potential role of DNA methylation via a high maternal folic acid intake in ASD and ASD-related traits. It will be useful to measure the proportion of variance in autism explained by maternal folic acid status after adujusting for other known risk factors, particularly, in women who take periconception folic acid. Lack of definitive biological markers of autism for a majority of cases makes it difficult to evaluate the proportion of risk attributable to maternal folic acid intake. Moreover, the level of maternal folic acid intake which may result in sufficient cause to contribute to development of various forms of autism is difficult to isolate. Several randomized clinical trials with folinic acid supplementation are in progress and may provide some answers about whether folic acid supplementation has a protective or adverse role, if any, in relation to development of ASD.

REFERENCES

1. Frazier, T.W.; Youngstrom, E.A.; Speer, L.; Embacher, R.; Law, P.; Constantino, J. Validation of proposed DSM-5 criteria for autism spectrum disorder. J. Am. Acad. Child Adol. Psychiatry 2012, 51, 28–40, doi:10.1016/j.jaac.2011.09.021.
2. Durkin, S.A.; Mathew, J.; Maenner, C.J.; Newschaffer, L.L.; Christopher, M.C.; Daniels, L. Advanced parental age and risk of autism spectrum disorder. Am. J. Epidemiol. 2006, 168, 1268–1276, doi:10.1093/aje/kwn250.
3. Jon, B. Prevalence of autism spectrum disorder: Autism developmental disabilities monitoring network, 14 Sites, United States, 2008. Morb. Mortal. Wkly. Rep. 2012, 61, 1–19.
4. Jon, B. Prevalence of autism spectrum disorder: Autism developmental disabilities monitoring network, 14 sites, United States, 2010. Morb. Mortal. Wkly. Rep. 2014, 63, 1–14.
5. James, S.J.; Melnyk, S.; Jernigan, S.; Cleves, M.A.; Halsted, C.H.; Wong, D.H.; Cutler, P.; Bock, K.; Boris, M.; Bradstreet, J.J.; et al. Metabolic endophenotype and related genotypes are associated with oxidative stress in children with autism. Neuropsychiatr. Genet. 2006, 141B, 947–956.
6. Lyall, K.; Schmidt, R.J.; Hertz-Picciotto, I. Maternal lifestyle and environmental risk factors for autism spectrum disorders. Int. J. Epidemiol. 2014, 43, 443–464, doi:10.1093/ije/dyt282.
7. Newschaffer, C.J.; Coren, L.A.; Daniels, J.; Giarelli, E.; Gerther, J.K.; Levy, S.E.; Mandell, D.S.; Miller, L.A.; Pinto-Martin, J.; Reaven, J.; et al. The epidemiology of autism spectrum-disorders. Annu. Rev. Publ. Health 2007, 28, 235–258, doi:10.1146/annurev.publhealth.28.021406.144007.
8. Suren, P.; Roth, C.; Bresnahan, M.; Haugen, M.; Hornig, M.; Hirtz, D.; Lie, K.K.; Lipkin, W.I.; Magnus, P.; Reichborn-Kjennerud, T.; et al. Association between maternal use of folic acid supplements and risk of autism spectrum disorders in children. JAMA 2013, 309, 570–577, doi:10.1001/jama.2012.155925.
9. Schmidt, R.J.; Tancredi, D.J.; Ozonoff, S.; Hansen, R.L.; Hartiala, J.; Allayee, H.; Schmidt, L.C.; Tassone, F.; Herzt-Picciotto, I. Maternal periconceptional folic acid intake and risk of autism spectrum disorders and developmental delay in

the CHARGE case-control study. Am. J. Clin. Nutr. 2012, 96, 80–89, doi:10.3945/ajcn.110.004416.

10. Berry, R.J.; Crider, K.S.; Yeargin-Allsopp, M. Periconceptional folic acid and risk of autism spectrum disorders. JAMA 2013, 309, 611–613, doi:10.1001/jama.2013.198.

11. Rogers, E.J. Has enhanced folate status during pregnancy altered natural selection and possibly autism prevalence? A closer look. Med. Hypotheses 2008, 71, 406–410, doi:10.1016/j.mehy.2008.04.013.

12. Friso, S.; Choi, S-W. Gene-nutrient interaction in one-carbon metabolism. Curr. Drug Metab. 2005, 6, 37–46, doi:10.2174/1389200052997339.

13. Junaid, M.A.; Kuizon, S.; Cardona, J.; Azher, T.; Murakami, N.; Pullarkat, R.K.; Brown, T.W. Folic acid supplementation dysregulates gene expression in lymphoblastoid cells—Implications in nutrition. Biochem. Biophys. Res. Commun. 2011, 412, 688–692, doi:10.1016/j.bbrc.2011.08.027.

14. Leeming, R.L.; Lucock, M. Autism: Is there a folate connection? J. Inherit. Metab. Dis. 2009, 32, 400–402, doi:10.1007/s10545-009-1093-0.

15. Bailey, L.B.; Caudill, M.A. Folate. In Present Knowledge in Nutrition, 10th ed.; Erdman, J.W., Macdonald, I.A., Zeisel, S.H., Eds.; Wiley-Blackwell: Ames, IA, USA, 2012; pp. 321–342.

16. Balion, C.; Kapur, B.M. Folate. Clinical utility of serum and red blood cell analysis. Clin. Lab. News 2011, 37, 8–10.

17. Baura, S.; Kuizon, S.; Chadman, K.K.; Flory, M.J.; Brown, W.T.; Junaid, M.A. Single base resolution of mouse offspring brain methylome reveals epigenome modifications caused by gestational folic acid. Epigenetics Chromatin 2014, 7, 1–15, doi:10.1186/1756-8935-7-1.

18. Main, P.A.; Angley, M.T.; Thomas, P.; O'Doherty, C.E.; Fenech, M. Folate and methionine metabolism in autism: A systematic review. Am. J. Clin. Nutr. 2010, 91, 1598–1620, doi:10.3945/ajcn.2009.29002.

19. Pu, D.; Shen, Y.; Wu, J. Association between MTHFR gene polymorphism and risk of autism spectrum disorder: A meta-analysis. Autism Res. 2013, 6, 384–392, doi:10.1002/aur.1300.

20. Wong, C.C.; Meaburn, E.L.; Ronald, A.; Price, T.S.; Jeffries, A.R.; Scalkwyk, L.C.; Plomin, R.; Mills, J. Methyomic analysis of monozygotic twins discordant for autism spectrum disorder and related behavioral traits. Mol. Psychiatry 2014, 19, 495–503, doi:10.1038/mp.2013.41.

21. Beard, C.M.; Panser, L.A.; Katusic, S.K. Is excess folic acid supplementation a risk factor for autism? Med. Hypotheses 2011, 77, 15–17, doi:10.1016/j.mehy.2011.03.013.

22. Smith, D.A.; Kim, Y.-I.; Refsum, H. Is folic acid good for everyone? Am. J. Clin. Nutr. 2008, 87, 517–533.

23. Amaral, D.G.; Schuman, C.M.; Nordhal, C.V. Neuroanatomy of autism. Trends Neurosci. 2008, 31, 137–145, doi:10.1016/j.tins.2007.12.005.

24. Courchense, E.; Pierce, K. Why the frontal cortex in autism might be talking only to itself: Local over-connectivity but long distance disconnection. Curr. Opin. Neurobiol. 2005, 15, 225–230, doi:10.1016/j.conb.2005.03.001.

25. Berry, R.J.; Li, Z.; Erickson, J.D. Prevention of neural tube defects in China. China–U.S. Collaborative Project of Neural Tube Prevention. N. Engl. J. Med. 1999, 341, 1485–1490, doi:10.1056/NEJM199911113412001.

26. Wolff, T.; Witkop, C.T.; Miller, T.; Syed, S.B. Folic acid supplementation for prevention of neural tube defects: An update of the evidence for the U.S. Preventive Services Task Force. Ann. Int. Med. 2009, 150, 632–639, doi:10.7326/0003-4819-150-9-200905050-00010.

27. Rodier, P.M.; Ingram, J.L.; Tisdale, B. Embryological origin for autism: Developmental anomalies of the cranial nerve motor nuclei. J. Comp. Neurol. 1996, 370, 247–261, doi:10.1002/(SICI)1096-9861(19960624)370:2<247::AID-CNE8>3.0.CO;2-2.

28. Reye-Engel, A.; Monoz, E.; Gaitan, M.J. Implications of human fertility of the 677C→T and 128A →C polymorphism of the MTHFR gene: Consequences of a possible genetic selection. Mol. Hum. Reprod. 2002, 8, 952–957, doi:10.1093/molehr/8.10.952.

29. Haggerty, P.; Campbell, D.M.; Duthie, S.; Andrews, K.; Hoad, G.; Piyathilake, C.; Fraser, I.; McNeill, G. Folic acid in pregnancy and embryo selection. BJOG 2008, 115, 851–856, doi:10.1111/j.1471-0528.2008.01737.x.

30. Nelen, W.L.D.M.; Blom, H.J.; Thomas, C.M.G.; Steegers, E.A.P.; Boers, G.H.J.; Eskes, T.K.A.B. Methylenetetrahydrofolate reductase polymorphism affects the change in homocysteine and folate concentration resulting from low dose of folic acid supplementation in women with unexplained recurrent miscarriages. J. Nutr. 1998, 128, 1336–1341.

31. Bailey, S.W.; Ayling, J.E. The extremely slow and variable activity of dihydrofolate reductase in human liver and its implications for high folic acid intake. Proc. Natl. Acad. Sci. USA 2009, 106, 15424–15429, doi:10.1073/pnas.0902072106.

32. Nurk, E.; Tell, G.S.; Refsum, H.; Ueland, P.M.; Vollset, S.E. Association between maternal methylenetetrahrdrofolate reductase polymorphism and adverse outcomes in pregnancy: The Hordaland homocysteine study. Am. J. Med. 2004, 117, 26–31, doi:10.1016/j.amjmed.2004.01.019.

33. Boris, M.D.; Goldblatt, P.A.; Galan, J.; James, J.S. Association of MTHFR gene variants and autism. J. Am. Phys. Surg. 2004, 9, 106–108.

34. Ramaekers, V.T.; Blau, N.; Sequeira, J.M.; Nassogne, M.C.; Quadros, E.V. Folate receptor autoimmunity and cerebral folate deficiency in low functioning autism with neurological defects. Neropediatrics 2007, 38, 276–281, doi:10.1055/s-2008-1065354.

35. Ramaekers, V.T.; Quadoros, E.V.; Sequeira, J.M. Role of receptor autoantibodies in infantile autism. Mol. Psychiatry 2013, 18, 270–271, doi:10.1038/mp.2012.22.

36. Frye, R.E.; Sequeira, J.M.; Quadros, E.V.; James, S.J.; Rossignol, D.A. Cerebral folate receptor autoantibodies in autism spectrum disorder. Mol. Psychiatry 2013, 18, 369–381, doi:10.1038/mp.2011.175.

37. Adams, B.J.; Audhya, T.; McDonough-Means, S.; Rubin, R.A.; Quig, D.; Geis, E.; Loresto, M.; Mitchell, J.; Atwood, S.; et al. Nutritional and metabolic status of children with autism vs. neurotypical children, and the association with autism severity. Nutr. MeTab. 2011, 8, 34–66, doi:10.1186/1743-7075-8-34.

38. Veena, S.R.; Krishnamachari, G.H.; Srinivasan, K.; Will, A.K.; Muthayya, S.; Kurpad, A.V.; Yajnik, C.S.; Fall, C.H.D. Higher maternal folate but not B12 concen-

trations during pregnancy are associated with better cognitive functional scores in 9–10 years old children in South India. J. Nutr. 2010, 140, 1014–1022, doi:10.3945/jn.109.118075.

39. Schmidt, R.J.; Hansen, R.L.; Hartiala, J.; Allayee, H.; Schmidt, L.C.; Tancredi, D.J.; Tassone, F.; Hertz-Piccott, I. Prenatal vitamins, one carbon metabolism gene variants, and risk of autism. Epidemiology 2011, 22, 476–485, doi:10.1097/EDE.0b013e31821d0e30.

40. James, S.J.; Cutler, P.; Melnyk, S.; Jernigan, S.; Janak, L.; Gaylor, D.W.; Neubrander, J.A. Metabolic biomarkers of increased oxidative stress and impaired methylation capacity in children with autism. Am. J. Clin. Nutr. 2004, 80, 1611–1617.

41. James, S.J.; Melnyk, S.; Fuchs, G.; Reid, T.; Jernigan, S.; Pavliv, O.; Hubanks, A.; Gaylor, D.W. Efficacy of methylcobalamin and folinic acid treatment on glutathione redox status in children with autism. Am. J. Clin. Nutr. 2009, 89, 425–430, doi:10.3945/ajcn.2008.26615.

42. James, S.J.; Melnyk, S.; Jernigan, S.; Pavliv, O.; Trusty, T.; Lehman, S.; Seidel, L.; Gaylor, D.W.; Cleves, M.A. A functional polymorphism in the reduced folate carrier gene and DNA hypomethylation in mothers of children with autism. Am. J. Med. Genet B Neuropsychiatr. Genet. 2010, 153B, 1209–1220.

43. Beaudet, A.L. Neuroscience: Preventable form of autism? Science 2012, 338, 342–343, doi:10.1126/science.1229178.

44. Neggers, Y.H. Increasing prevalence, changes in diagnostic criteria, and nutritional risk factors for autism spectrum disorders. ISRN Nutr. 2014, doi:10.115/2014/514026.

45. Frye, R.E. A folinic acid intervention for ASD: Links to folate receptor-alpha auto-immunity and redox metabolism. Available online: http://clinicaltrials.gov/c2/show/NCT01602016 (accessed on 15 September 2014).

46. Follow-up study of late effects of periconceptional folic acid in mothers and off-springs in the community intervention program population: The Chinese and families study. Available online: http://clinicaltrials.gov/show/NCT01365975 (accessed on 15 September 2014).

There is one table that is not available in this version of the article. To view this additional information, please use the citation on the first page of this chapter.

PART IV

MACRONUTRIENT COMPOSITION, ENERGETICS, AND ENERGY BALANCE

CHAPTER 9

Consumption of Fructose-Sweetened Beverages for 10 Weeks Reduces Net Fat Oxidation and Energy Expenditure in Overweight/Obese Men and Women

C. L. COX, K. L. STANHOPE, J. M. SCHWARZ, J. L. GRAHAM, B. HATCHER, S. C. GRIFFEN, A. A. BREMER, L. BERGLUND, J. P. MCGAHAN, P. J. HAVEL, AND N. L. KEIM

9.1 INTRODUCTION

An increase in the use of sweeteners containing fructose has occurred in parallel with the increasing prevalence of overweight and obesity over the past three decades in the United States (Bray et al., 2004), suggesting that increased consumption of fructose, high fructose corn syrup and/ or sucrose may contribute to the current epidemic of obesity and the increased incidence of metabolic syndrome (MetSyn) (Bray et al., 2004; Havel, 2005). In animal studies, consumption of diets high in fructose produces obesity, insulin resistance and dyslipidemia (Storlien et al., 1993; Martinez et al., 1994; Okazaki et al., 1994; Bezerra et al., 2000;

Consumption of Fructose-Sweetened Beverages for 10 Weeks Reduces Net Fat Oxidation and Energy Expenditure in Overweight/Obese Men and Women. Cox CL, Stanhope KL, Schwarz JM, Graham JL, Hatcher B, Griffen SC, Bremer AA, Berglund L, McGahan JP, Havel PJ, and Keim NL. European Journal of Clinical Nutrition *66 (2012). doi:10.1038/ejcn.2011.159. Reprinted with permission from the authors.*

Elliott et al., 2002; Havel, 2005). In humans, moderate fructose consumption has no apparent health concerns (Dolan et al., 2010; Rizkalla, 2010), but the health consequences of fructose consumption in large amounts are less clear, particularly its effects on substrate utilization and body weight regulation.

Recently we reported that consumption of fructose-sweetened beverages for 10 weeks, at 25% of energy requirements, increased hepatic de novo lipogenesis (DNL), promoted accumulation of intra-abdominal fat, produced a more atherogenic lipid profile and reduced insulin sensitivity in older, overweight and obese adults compared with isocaloric consumption of glucose (Stanhope et al., 2009). We hypothesized that the increased rate of DNL following consumption of fructose leads to an accumulation of hepatic lipid, which promotes dyslipidemia and decreases in insulin sensitivity (Stanhope et al., 2009). McGarry (1995) observed that increases of DNL led to a concomitant reduction in fat oxidation, which may contribute to an accumulation of hepatic lipid. Therefore, to determine if the increases of DNL associated with increased fructose consumption were accompanied by decreased fat oxidation, we measured resting and postprandial substrate utilization and energy expenditure using indirect calorimetry in the overweight and obese adults during their participation in the study mentioned above (Stanhope et al., 2009).

Previous investigations of the effects of fructose consumption on substrate utilization and energy expenditure are limited to acute or short-term studies (Tappy et al., 1986; Schwarz et al., 1989, 1992a, 1992b; Markov et al., 2000; Chong et al., 2007; Couchepin et al., 2008). Our results from this 10 week study indicate that in overweight/obese adults, 40–72 years of age, sustained consumption of fructose-sweetened beverages, at 25% of energy requirements, leads to decreased net postprandial fat oxidation and increased net postprandial carbohydrate oxidation, similar to what has been observed in short-term studies. In addition, we found that resting energy expenditure (REE) decreased significantly after 10 weeks in subjects consuming fructose-sweetened beverages, despite increases of body weight.

9.2 SUBJECTS AND METHODS

9.2.1 STUDY DESIGN

This was a parallel arm study with three phases: (1) a 2-week inpatient baseline period; (2) an 8-week outpatient intervention period and (3) a 2-week inpatient intervention period.

The details of the study design and diet interventions have been described (Stanhope et al., 2009). Briefly, during baseline, subjects resided as inpatients and consumed an energy-balanced diet. Procedures included the following: indirect calorimetry, dual energy X-ray absorptiometry and a computerized tomography scan of the abdomen. Subjects then began the 8-week outpatient intervention and consumed either fructose- or glucose-sweetened beverages at 25% of energy requirements with self-selected ad libitum diets. Subjects returned as inpatients for the final 2 weeks of intervention, during which the procedures were repeated while the subjects consumed their assigned glucose- or fructose-sweetened beverages as part of an energy-balanced diet.

9.2.2 SUBJECTS

Participants were recruited through advertisements and underwent a telephone and an in-person interview to assess eligibility. Inclusion criteria included age 40–72 years, body mass index 25–35 kg/m^2 and stable body weight during the prior 6 months. Women were post-menopausal. Exclusion criteria included the following: evidence of diabetes, renal or hepatic disease, fasting serum TG concentrations >400 mg/dl, blood pressure >140/90 mm Hg and surgery for weight loss. Individuals who smoked, reported exercise of more than 3.5 h/week at a level more vigorous than walking, or having used thyroid, lipid-lowering, glucose-lowering, anti-hypertensive, anti-depressant or weight loss medications were also excluded. Diet-related exclusion criteria included habitual ingestion of more than one sugar-sweetened beverage/day or more than two alcoholic beverages/day. The UC-Davis Institutional Review Board approved the study, and subjects provided informed consent to participate. Initially 39 subjects enrolled, but seven subjects did not complete the study due to personal or work-related conflicts, and one subject opted out of the indirect calorimetry protocol. Thus, a total of 31 subjects completed the study: n=15 for the glucose group and n=16 for the fructose group. As previously reported (Stanhope et al., 2009), baseline characteristics between the two experimental sugar groups were not different (Table 1).

TABLE 1: Subject characteristics at baseline.[a,b]

Parameter	Glucose		Fructose	
	Male	Female	Male	Female
	(n=7)	(n=8)	(n=9)	(n=7)
Age (years)	54±3	56±2	52±4	53±3
Weight (kg)	88.4±2.9	84.0±4.5	89.3±2.9	80.5±4.6
BMI (kg/m²)	29.3±1.1	29.4±1.3	28.4±0.7	30.0±1.2
Body fat (%)	29.4±1.1	43.2±1.5	28.5±1.3	40.5±2.1
Fat-free mass (kg)	63.3±1.4	48.2±2.1	64.8±2.0	47.6±2.4
Triglycerides (mg/dl)	148±31	145±23	131±21	151±36
Cholesterol (mg/dl)	179±14	193±10	176±6	197±14
LDL-cholesterol (mg/dl)	124±5	123±11	107±7	125±10
Glucose (mg/dl)	89±2	89±3	88±1	90±2

Abbreviations: BMI, body mass index; LDL, low-density lipoprotein. [a]Values are mean±s.e.m. Clinical chemistry values are fasting values. There were no significant differences between sugar × gender groups. [b]Data were published previously (Stanhope et al., 2009).

9.2.3 DIETS

During the inpatient metabolic phases, subjects consumed diets designed to maintain energy balance providing 15% of energy as protein, 30% as fat and 55% as carbohydrate. Daily energy intake was calculated at baseline using the Mifflin equation to estimate REE (Mifflin et al., 1990) and adjusted for activity using a multiplication factor of 1.5. During baseline, the carbohydrate content consisted primarily of complex carbohydrates and contained 8.8 ± 1.2 g of dietary fiber/1000 kcal. For the final 2-week inpatient intervention period, subjects consumed diets at the baseline energy level and macronutrient composition except that 30% of energy was from complex carbohydrates and 25% was provided by fructose- or glucose-sweetened beverages. Additional details about the diet intake for inpatient and outpatient phases have been described previously (Stanhope et al., 2009).

9.2.4 INDIRECT CALORIMETRY

An automated metabolic measuring cart (Truemax 2400 Metabolic Measurement System, Parvomedics, Salt Lake City, UT, USA) was used to measure rates of O_2 consumption (VO_2) and CO_2 production (VCO_2). Gas analyzers were calibrated using a certified gas mixture of known O_2 and CO_2 concentrations, and the flowmeter was calibrated using a 3 l syringe, four times daily (0700, 1100, 1500 and 1800 hours). The protocol was conducted at the CCRC at baseline-week 0, and at intervention-week 10 and was preceded by a minimum of 5 days of controlled diet. REE was measured on two separate days during week 0 and again at week 10, whereas postprandial energy expenditure (PPEE) was measured only once during week 0 and once during week 10, following a REE measurement. Respiratory gases were collected while subjects were in a semi-reclined position and wore a facemask fitted securely, covering the nose and mouth. The facemask was attached to tubing connected to the cart's mixing chamber. Through the facemask, subjects inhaled room air, and all expired breath was trapped by the mask and directed to the cart's mixing chamber for volume and gas analyses. Prior to the test, subjects fasted overnight for 13.5 h and rested quietly for at least 10 min before measurements commenced. REE was measured at 0730 and 0830 hours for 20 min periods. For PPEE, respiratory gases were collected for 15 min every hour over the next 14 h; the most stable 10-min interval of each 15-min collection was selected to represent PPEE for that hour. During the protocol subjects consumed the controlled breakfast (0900 hours), lunch (1300 hours) and dinner (1800 hours) meals. They were permitted to perform light activities associated with living in a metabolic ward, but rested in a semi-reclined position for at least 5 min before each measurement. Indirect calorimetry data were analyzed for all 31 subjects for REE, but only for 30 subjects for PPEE (n=15 in the fructose group and n=15 in the glucose group) due to procedure scheduling conflicts for one female in the fructose group.

9.2.5 CALCULATION OF ENERGY EXPENDITURE AND SUBSTRATE OXIDATION RATES

Energy expenditure was calculated using the Weir equation (Weir, 1990):

$$Kcal/min = (3.941 \times VO_2) + (1.106 \times VCO_2) - (2.17 \times urinary\ N)$$

Net carbohydrate oxidation (CHO-Ox) and net fat oxidation (FAT-Ox) were calculated using the following formulas (Frayn, 1983):

$$Fat - Ox\ (g/min) = (1.67 \times VO_2) - (1.67 \times VCO_2) - (1.92 \times urinary\ N)$$

$$CHO \times Ox\ (g/min) = (4.55 \times VCO_2) - (3.21 \times VO_2) - (2.87 \times urinary\ N)$$

For all equations, units for VO_2 and VCO_2 are in l/min. To estimate urinary nitrogen excretion, in g/min, a constant rate of protein catabolism was assumed, equivalent to the 24-h protein intake, as reported by Bingham (2003).

9.2.6 MEASUREMENTS OF BODY COMPOSITION

Total body fat and fat free mass were determined by dual energy X-ray absorptiometry, and intra- and extra-abdominal fat were measured by computerized tomography scan as described previously (Stanhope et al., 2009).

9.2.7 DATA ANALYSIS

REE, PPEE and corresponding net substrate oxidation values for baseline and intervention were calculated using minute-by-minute values for VO_2 and VCO_2 (l/min). Overall PPEE and postprandial substrate oxidation rates were estimated by averaging values for the 14 postprandial time periods. Statistical tests were performed with SAS 9.2 (SAS Institute, Cary, NC, USA). The absolute or percent change for each outcome was analyzed in a three-factor (type of sugar, gender, and + or −MetSyn) mixed procedure (PROC MIXED) analysis of variance model with adjustment for the change in fat-free mass. The model was also run using baseline REE and PPEE as covariates, and the changes in REE and PPEE were still significant and comparable to those obtained when the model was adjusted for fat-free mass. MetSyn was defined as having at least three MetSyn risk factors as defined by the American Heart Association/National Heart Lung and Blood Institute (Grundy et al., 2004). There were 9 subjects with Met-Syn (fructose n=5, glucose n=4) and 22 subjects without MetSyn (fructose n=11, glucose n=11). Outcomes with least squares means of the change (10 week versus 0 week) significantly different than zero were identified. Statistical tests with P-values <0.05 were considered significant. Data are presented as mean±s.e.m.

9.3 RESULTS

9.3.1 BODY WEIGHT AND COMPOSITION

As reported previously (Stanhope et al., 2009), despite comparable weight gain (~1–2% of initial body weight) during the 8-week outpatient intervention, subjects consuming fructose primarily exhibited increases of visceral adipose tissue, whereas in subjects consuming glucose subcutaneous adipose tissue was preferentially increased.

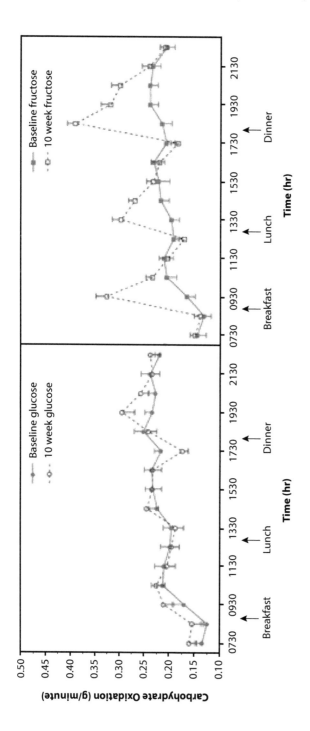

FIGURE 1: Net carbohydrate oxidation rate (g/min) profiles over 15 h for subjects consuming glucose- and fructose-sweetened beverages. Subjects consumed meals at 0900, 1300 and 1800 hours as indicated. The first two data points represent resting values and the remaining 14 data points represent postprandial values. Data points represent the mean of 10-min measurements±s.e.m. with n=31 (fructose group n=16; glucose group n=15) for resting values and n=30 (fructose group n=15; glucose group n=15) for postprandial values.

TABLE 2: Net carbohydrate and fat oxidation rates and percent change after consumption of glucose- or fructose-sweetened beverages for 10 weeks.

Sample		Fructose baseline	Fructose week 10	Fructose % change	Glucose baseline	Glucose week 10	Glucose % change	Sugar[b]	Sugar × MetSyn[b]
Carbohydrate oxidation rate, g/min									
Fasting	All	0.16±0.06	0.17±0.01	14.8±17.6	0.16±0.01	0.18±0.07	23.6±11.5	0.745	0.047
Fasting	(−)MetSyn	0.16±0.02	0.15±0.01	−2.2±8.5	0.16±0.02	0.20±0.02	30.4±14.7		
Fasting	(+)MetSyn	0.17±0.04	0.20±0.03	52.0±52.9*	0.14±0.02	0.14±0.02	5.2±13.9		
Postprandial	All	0.24±0.02	0.29±0.01	23.5±8.6***	0.25±0.01	0.26±0.01	3.8±3.3	0.005	
Postprandial	(−)MetSyn	0.26±0.02	0.28±0.01	11.9±4.8	0.25±0.02	0.26±0.02	5.2±4.2		
Postprandial	(+)MetSyn	0.21±0.03	0.29±0.02	46.6±21.8***	0.23±0.02	0.23±0.02	−0.04±8.3	0.012	
Fat oxidation rate, g/min									
Fasting	All	0.06±0.01	0.05±0.01	−13.1±8.0	0.06±0.01	0.05±0.01	−14.2±14.1	0.366	0.037
Fasting	(−)MetSyn	0.06±0.01	0.05±0.01	−7.0±7.6	0.06±0.01	0.05±0.01	−26.7±15.5*		
Fasting	(+)MetSyn	0.06±0.01	0.04±0.03	−26.5±19.8	0.05±0.01	0.05±0.01	20.3±27.4		
Postprandial	All	0.05±0.01	0.03±0.01	−38.2±4.8***	0.05±0.01	0.04±0.01	−8.6±6.1	0.001	0.020
Postprandial	(−)MetSyn	0.05±0.01	0.03±0.01	−33.7±4.3***	0.05±0.01	0.04±0.01	−14.1±7.4*		
Postprandial	(+)MetSyn	0.06±0.01	0.03±0.01	−47.4±11.0***	0.04±0.01	0.04±0.01	6.3±6.9		

Abbreviations: MetSyn, metabolic syndrome; (−), without; (+), with. [a] Values are mean±s.e.m. Sample sizes in sugar × MetSyn groups are as follows: fructose(−)MetSyn, n=11 resting, n=10 postprandial; fructose(+)MetSyn, n=5; glucose(−)MetSyn, n=11; glucose(+)MetSyn, n=4. [b] PROC MIXED three-factor analysis of variance (sugar, gender, (+) or (−) MetSyn) adjusted for change in fat-free mass. $*P<0.05$, $***P<0.001$ for changes significantly different from zero.

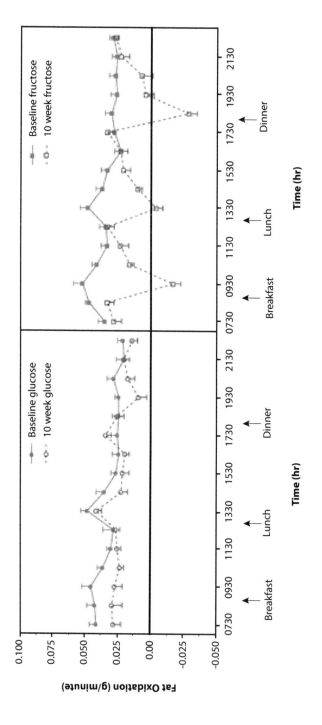

FIGURE 2: Net fat oxidation rate (g/min) profiles over 15h for subjects consuming glucose- and fructose-sweetened beverages. Subjects consumed meals at 0900, 1300 and 1800 hours as indicated. The first two data points represent resting values and the remaining 14 data points represent postprandial values. Data points represent the mean of 10-min measurements±s.e.m. with n=31 (fructose group n=16; glucose group n=15) for resting values and n=30 (fructose group n=15; glucose group n=15) for postprandial values.

9.3.2 NET SUBSTRATE OXIDATION RATES

Overall, fasting CHO-Ox did not change in response to glucose (P=0.29) or fructose (P=0.11) consumption, however subjects with MetSyn consuming fructose exhibited significant increases of fasting CHO-Ox (P=0.02) after 10 weeks of intervention (Table 2). Postprandial CHO-Ox increased significantly from baseline in subjects consuming fructose-sweetened beverages (P<0.0001), but not in those consuming glucose-sweetened beverages (P=0.54) (Table 2, Figure 1). Similar to the fasting condition, subjects with MetSyn consuming fructose exhibited marked percent increases of postprandial CHO-ox (P<0.0001), whereas in subjects consuming fructose without MetSyn postprandial CHO-ox rates were not changed from baseline values (P=0.34) (sugar × MetSyn interaction: P=0.01). There was also an effect of gender on the change of postprandial CHO-Ox (P=0.05), such that male subjects in both sugar groups exhibited greater increases (21.35±9.8%; P<0.001) than female subjects (5.75±3.9%; P=0.81).

The change of fasting FAT-Ox differed both by sugar and by the presence/absence of MetSyn. While fasting FAT-Ox tended to decrease in the fructose group at 10 weeks (P=0.15), this trend was driven by the larger decreases that were observed in subjects with MetSyn (P=0.14) as opposed to those without MetSyn (P=0.74). Although there was no overall change in fasting FAT-Ox in the glucose group at 10 weeks (P=0.92), there was a significant decrease in subjects without MetSyn (P<0.05) (Table 2). In subjects consuming fructose-sweetened beverages postprandial FAT-Ox rates decreased significantly both compared with baseline values (P<0.0001) and compared with subjects consuming glucose-sweetened beverages (P<0.0001) (Table 2, Figure 2). Overall the percent decrease of postprandial FAT-Ox was significant in subjects consuming fructose regardless of the presence (P<0.0001) or absence (P=0.0002) of MetSyn, whereas with glucose consumption only subjects without MetSyn exhibited statistically significant decreases of postprandial FAT-Ox (P=0.03) (Table 2).

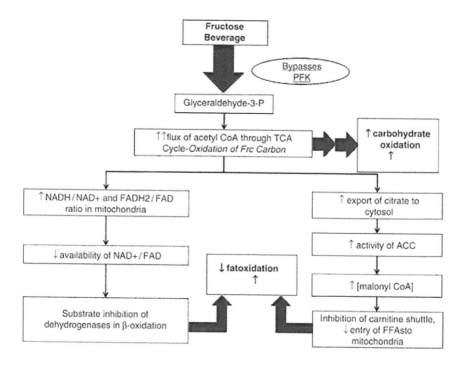

FIGURE 3: Proposed mechanisms contributing to observed changes of substrate utilization in subjects consuming fructose-sweetened beverages. In the liver fructose is phosphorylated by fructokinase (which is not regulated by cellular energy status) and largely bypasses phosphofructokinase (PFK), the enzyme catalyzing the rate-limiting step of glycolysis (which is subject to inhibition by ATP and citrate). Ultimately fructose enters the glycolytic pathway as glyceraldehyde-3-phosphate. Following a high-fructose meal, an unregulated flux of fructose (Frc) carbon upregulates carbohydrate metabolism in the liver (increased CHO-Ox), leading to an increased flux of acetyl CoA through the tricarboxylic acid (TCA) cycle and a concomitant increase in cellular energy status (increased ATP/ADP ratio and NADH/NAD+ ratio). A high NADH/NAD+ ratio in the mitochondria results in substrate inhibition of isocitrate dehydrogenase in the TCA cycle, leading to increased export of citrate to the cytosol, activation of acetyl-CoA carboxylase (ACC), and increased production of malonyl-CoA, the precursor to fatty acid synthesis (DNL). Elevated cytosolic concentrations of malonyl-CoA inhibit the carnitine shuttle via carnitine palmitoyl transferase, leading to reduced entry of fatty acids into the mitochondria, and decreased fat oxidation. The elevation of cellular energy status following a high-fructose meal would also lead to reduced mitochondrial availability of the fixed pool of oxidized cofactors NAD+ and FAD, which are required substrates for β-oxidation, also resulting in reduced fat oxidation (Williamson and Cooper, 1980; Mayes, 1993; McGarry, 1995; Locke et al., 2008).

9.3.3 ENERGY EXPENDITURE

REE was significantly decreased from baseline values by week 10 in subjects consuming fructose (P=0.03) but not in those consuming glucose (P=0.86). PPEE also tended to decrease from baseline values in subjects in the fructose group but the change was not statistically significant (P=0.19). PPEE was unchanged from baseline values in the glucose group (P=0.86) (Table 3).

TABLE 3: Energy expenditure before and after consumption of glucose- or fructose-sweetened beverages for 10 weeks.

	Fructose Baseline	Fructose 10 weeks	Fructose change	Glucose Baseline	Glucose 10 weeks	Glucose change	P-value for effect of sugar[b]
Energy expenditure (kcal/min)							
Resting[b]	1.19±0.06	1.10±0.04	−0.09 ± 0.04*	1.17±0.07	1.15±0.05	−0.02 ± 0.04	0.108
Post-prandial	1.41±0.06	1.37±0.05	−0.05 ± 0.02	1.40±0.06	1.36±0.05	−0.03 ± 0.03	0.445

[a] Values are means±s.e.m. Fasting values are based on n=31 (fructose group n=16; glucose group n=15) and postprandial values are based on n=30 (fructose group n=15; glucose group n=15). [b] PROC MIXED three-way model (sugar, gender, (+) or (−) metabolic syndrome) adjusted for change in fat-free mass. *P<0.05 for changes significantly different from zero.

9.4 DISCUSSION

9.4.1 NET SUBSTRATE OXIDATION RATES

The results of acute and short-term studies of fructose ingestion, ranging from periods of 4 h to 6 days, indicate that when fructose is consumed in large amounts ranging from 30–50% of total calories, net fat oxidation is decreased and net carbohydrate oxidation is increased (Tappy et al.,

1986; Schwarz et al., 1989, 1992a, 1992b; Markov et al., 2000; Chong et al., 2007; Couchepin et al., 2008). However, the majority of these studies only examined the effects of consuming a single meal containing fructose, and those that examined the effects of multiple days of fructose consumption did not do so under energy-balanced conditions, but rather during consumption of 25–50% excess calories. Importantly, our results demonstrate that the acute effects of fructose consumption persist when fructose is consumed over longer periods (10 weeks), and in subjects consuming an energy-balanced diet, suggesting that fructose-induced changes of the regulation of key metabolic pathways involved in cellular energy utilization are sustained even in the absence of positive energy balance.

We have previously reported that sustained consumption of fructose-sweetened beverages increased hepatic DNL in these same subjects (Stanhope et al., 2009). Schwarz et al. (1995) reported strong correlations between changes of substrate oxidation rates (increased CHO-ox and decreased FAT-ox) and increases of DNL in subjects consuming 25 or 50% excess energy as carbohydrate (carbohydrate composition not specified) for 5 days. Here we demonstrate that, under energy-balanced conditions, consuming 25% of energy from fructose leads to reduced net fat oxidation and increased net carbohydrate oxidation, in addition to previously reported increases of DNL (Stanhope et al., 2009). This relationship was not observed with isocaloric glucose consumption suggesting that consumption of fructose, not carbohydrate in general (in this case monosaccharides), leads to these changes in the regulation of substrate oxidation and DNL; however, recent evidence suggests that other factors such as differences in the amylose/amylopectin ratio of carbohydrate rich foods may also lead to similar changes (Isken et al., 2010).

These findings support our hypothesis that decreases of fat oxidation occurring concurrently with fructose-induced upregulation of DNL promotes increases of hepatic lipid content, which may mediate the adverse changes in lipid metabolism and decreased insulin sensitivity we have reported previously (Stanhope et al., 2009). These results are also consistent with the mechanism proposed by McGarry and others by which consumption of fructose leads to reductions of fat oxidation and increased carbohydrate oxidation (Figure 3) (Mayes, 1993; McGarry, 1995). It must be emphasized that values for substrate utilization derived from indirect calorimetry represent rates of substrate disappearance that may not always equate with rates of substrate oxidation. Determination of actual rates of substrate oxidation would require additional studies using isotopic tracer methodology.

9.4.2 EFFECTS OF METSYN

We observed that subjects with MetSyn consuming fructose-sweetened beverages exhibited the largest decreases of postprandial fat oxidation rates and increases of carbohydrate oxidation rates (Table 2). This relationship was not evident in subjects consuming glucose. It should be noted that the number of subjects entering the study with MetSyn was small, five in the fructose group and four in the glucose group. Additionally, as we have reported previously (Stanhope et al., 2009), 10 weeks of fructose consumption promoted the development of risk factors for MetSyn, such as accumulation of intra-abdominal fat, dyslipidemia and insulin resistance. Hence, it is likely that we are observing a worsening of metabolic function in subjects consuming fructose that is further exacerbated in those who already had evidence of MetSyn before the intervention.

9.4.3 EFFECTS OF GENDER

We also observed that men consuming both glucose- and fructose-sweetened beverages exhibited greater increases of postprandial carbohydrate oxidation than women. These findings support those of Couchepin et al. (2008) who reported a significant increase of carbohydrate oxidation in male, but not female subjects consuming fructose. Together these findings suggest that there is a gender-specific response with respect to changes of substrate utilization following sustained consumption of fructose.

9.4.4 ENERGY EXPENDITURE

The decrease of REE that we observed in subjects consuming fructose was unexpected and conflicts with the findings of several previous short-term studies that reported increased energy expenditure following administration of oral fructose as compared with consumption of glucose (Tappy et al., 1986; Schwarz et al., 1989, 1992a, 1992b). These changes suggest that sustained fructose consumption may contribute to an overall reduction in energy expenditure, which could increase the risk for weight gain if energy intake is not adjusted downward accordingly. For example, if the mean measured decrease of REE associated with 10 weeks of fructose consumption, 0.09 kcal/min, was maintained for 1 year it could total ~15 000 kcals, assuming that REE reflects metabolism during rest/sleep periods adding to about 8 h/d; potentially, a gain of ~1.6 kg of body fat could result. Additional studies examining the effects of chronic sugar consumption on 24-hour energy expenditure conducted in a whole-room calorimeter are needed to confirm these findings and determine if the observed reductions in metabolic rate are directly related to fructose or to sweetener (sucrose, high fructose corn syrup, and so on) consumption in general. We are currently performing such measurements.

9.5 CONCLUSIONS

Consumption of fructose at 25% of energy requirements for 10 weeks, when compared with isocaloric consumption of glucose, leads to significant reductions of net postprandial fat oxidation and increases of net postprandial carbohydrate oxidation. Furthermore, the results of this study demonstrate that these changes are evident even when fructose is consumed under energy-balanced conditions. We also report that REE is reduced compared with baseline values in subjects consuming fructose-sweetened beverages for 10 weeks. These findings may thus have important implications with regard to long-term energy balance in individuals consistently consuming large amounts of dietary fructose.

REFERENCES

1. Bezerra RM, Ueno M, Silva MS, Tavares DQ, Carvalho CR, Saad MJ (2000). A high fructose diet affects the early steps of insulin action in muscle and liver of rats. J Nutr 130, 1531–1535.

2. Bingham SA (2003). Urine nitrogen as a biomarker for the validation of dietary protein intake. J Nutr 133(Suppl 3), 921S–924S.
3. Bray GA, Nielsen SJ, Popkin BM (2004). Consumption of high-fructose corn syrup in beverages may play a role in the epidemic of obesity. Am J Clin Nutr 79, 537–543.
4. Chong MF, Fielding BA, Frayn KN (2007). Mechanisms for the acute effect of fructose on postprandial lipemia. Am J Clin Nutr 85, 1511–1520.
5. Couchepin C, Le KA, Bortolotti M, da Encarnacao JA, Oboni JB, Tran C et al. (2008). Markedly blunted metabolic effects of fructose in healthy young female subjects compared with male subjects. Diabetes Care 31, 1254–1256.
6. Dolan LC, Potter SM, Burdock GA (2010). Evidence-based review on the effect of normal dietary consumption of fructose on blood lipids and body weight of overweight and obese individuals. Crit Rev Food Sci Nutr 50: 889–918.
7. Elliott SS, Keim NL, Stern JS, Teff K, Havel PJ (2002). Fructose, weight gain, and the insulin resistance syndrome. Am J Clin Nutr 76, 911–922.
8. Frayn KN (1983). Calculation of substrate oxidation rates in vivo from gaseous exchange. J Appl Physiol 55, 628–634.
9. Grundy SM, Brewer Jr HB, Cleeman JI, Smith Jr SC, Lenfant C, American Heart Association et al. (2004). Definition of metabolic syndrome: report of the National Heart, Lung, and Blood Institute/American Heart Association conference on scientific issues related to definition. Circulation 109: 433–438.
10. Havel PJ (2005). Dietary fructose: implications for dysregulation of energy homeostasis and lipid/carbohydrate metabolism. Nutr Rev 63, 133–157.
11. Isken F, Klaus S, Petzke KJ, Loddenkemper C, Pfeiffer AF, Weickert MO (2010). Impairment of fat oxidation under high- vs. low-glycemic index diet occurs before the development of an obese phenotype. Am J Physiol Endocrinol Metab 298, E287–E295.
12. Locke GA, Cheng D, Witmer MR, Tamura JK, Haque T, Carney RF et al. (2008). Differential activation of recombinant human acetyl-CoA carboxylases 1 and 2 by citrate. Arch Biochem Biophys 475, 72–79.
13. Markov AK, Neely WA, Didlake RH, Terry 3rd J, Causey A, Lehan PH (2000). Metabolic responses to fructose-1,6-diphosphate in healthy subjects. Metabolism 49, 698–703. ChemPort |
14. Martinez FJ, Rizza RA, Romero JC (1994). High-fructose feeding elicits insulin resistance, hyperinsulinism, and hypertension in normal mongrel dogs. Hypertension 23, 456–463.
15. Mayes PA (1993). Intermediary metabolism of fructose. Am J Clin Nutr 58, 754S–765S.
16. McGarry JD (1995). Malonyl-CoA and carnitine palmitoyltransferase I: an expanding partnership. Biochem Soc Trans 23, 481–485.
17. Mifflin MD, St Jeor ST, Hill LA, Scott BJ, Daugherty SA, Koh YO (1990). A new predictive equation for resting energy expenditure in healthy individuals. Am J Clin Nutr 51, 241–247.
18. Okazaki M, Zhang H, Yoshida Y, Ichino K, Nakayama S, Oguchi K (1994). Correlation between plasma fibrinogen and serum lipids in rats with hyperlipidemia induced by cholesterol free-high fructose or high cholesterol diet. J Nutr Sci Vitaminol (Tokyo) 40, 479–489.

19. Rizkalla SW (2010). Health implications of fructose consumption: A review of recent data. Nutr Metab (Lond) 7, 82–98.

20. Schwarz JM, Acheson KJ, Tappy L, Piolino V, Muller MJ, Felber JP et al. (1992a). Thermogenesis and fructose metabolism in humans. Am J Physiol 262, E591–E598.

21. Schwarz JM, Neese RA, Turner S, Dare D, Hellerstein MK (1995). Short-term alterations in carbohydrate energy intake in humans. Striking effects on hepatic glucose production, de novo lipogenesis, lipolysis, and whole-body fuel selection. J Clin Invest 96, 2735–2743. ChemPort |

22. Schwarz JM, Schutz Y, Froidevaux F, Acheson KJ, Jeanpretre N, Schneider H et al. (1989). Thermogenesis in men and women induced by fructose vs glucose added to a meal. Am J Clin Nutr 49, 667–674.

23. Schwarz JM, Schutz Y, Piolino V, Schneider H, Felber JP, Jequier E (1992b). Thermogenesis in obese women: effect of fructose vs. glucose added to a meal. Am J Physiol 262, E394–E401.

24. Stanhope KL, Schwarz JM, Keim NL, Griffen SC, Bremer AA, Graham JL et al. (2009). Consuming fructose-sweetened, not glucose-sweetened, beverages increases visceral adiposity and lipids and decreases insulin sensitivity in overweight/obese humans. J Clin Invest 119, 1322–1334. ChemPort |

25. Storlien LH, Oakes ND, Pan DA, Kusunoki M, Jenkins AB (1993). Syndromes of insulin resistance in the rat. Inducement by diet and amelioration with benfluorex. Diabetes 42, 457–462. ChemPort |

26. Tappy L, Randin JP, Felber JP, Chiolero R, Simonson DC, Jequier E et al. (1986). Comparison of thermogenic effect of fructose and glucose in normal humans. Am J Physiol 250, E718–E724.

27. Weir JB (1990). New methods for calculating metabolic rate with special reference to protein metabolism. 1949. Nutrition 6, 213–221.

28. Williamson JR, Cooper RH (1980). Regulation of the citric acid cycle in mammalian systems. FEBS Lett 117 Suppl, K73–K85.

CHAPTER 10

Effects of High Carbohydrate or High Protein Energy-Restricted Diets Combined With Resistance-Exercise on Weight Loss and Markers of Health in Women with Serum Triglyceride Levels Above or Below Median Values

JONATHAN M. OLIVER, JULIE Y. KRESTA, MIKE BYRD, CLAIRE CANON, MICHELLE MARDOCK, SUNDAY SIMBO, PETER JUNG, BRITTANIE LOCKARD, DEEPESH KHANNA, MAJID KOOZEHCHIAN, CHRIS RASMUSSEN, CHAD KERKSICK, AND RICHARD KREIDER

10.1 BACKGROUND

A diet high in protein has been shown to have beneficial effects on weight loss and triglyceride (TG) levels when combined with exercise. Recent research has also shown that a diet high in protein in the absence of exercise promotes more favorable results for individuals above the median TG (mTG) levels (>133 mg/dL). The purpose of this study was to determine if women with TG above median values experience greater benefits to a diet and circuit resistance-training program.

Effects of High Carbohydrate or High Protein Energy-Restricted Diets Combined With Resistance-Exercise on Weight Loss and Markers of Health in Women with Serum Triglyceride Levels Above or Below Median Values.© *Oliver JM et al.* Journal of the International Society of Sports Nutrition *7, Suppl 1 (2010), doi:10.1186/1550-2783-7-S1-P9. Licensed under a Creative Commons Attribution License.*

10.2 METHODS

442 apparently healthy sedentary obese women (48±12 yrs, 64±3 in, 201±39 lbs, 45±5 % fat) completed a 10-wk exercise and diet program. All subjects participated in Curves circuit training (30-minute hydraulic resistance exercise interspersed with recovery floor calisthenics performed at 30-seconed intervals 3 days/wk) and weight loss program (1,200 kcal/d for 1 wk; 1,600 kcal/d for 9 wks). Subjects were randomly assigned to a high protein or high carbohydrate isocaloric diet. The high protein (HP) group (n=200) consumed 30% fat, 55-63% protein, and 9-15% carbohydrate diet while the high carbohydrate (HC) group (n=242) consumed 30% fat, 55% carbohydrate, and 15% protein diet. Pre and post measurements included standard anthropometric measurements including dual energy X-ray absorptiometry (DEXA), as well as resting energy expenditure (REE), metabolic blood analysis, and blood pressure. Subjects were stratified into a lower or higher TG group based on the mTG value observed (125 mg/dL). Data were analyzed by MANOVA with repeated measures and are presented as means ± SD percent changes from baseline.

10.3 RESULTS

Fasting serum TG levels differed between groups stratified based on mTG levels (<mTG 86±24 vs >mTG 204±84 mg/dL, p=0.001). Time effects were observed in all anthropometric measurements including waist and hip, as well as weight loss, fat mass and percent body fat. Subjects on the HP diet experienced greater reductions in weight than those on the HC diet (HP -3.1±3.4%; HC -2.3±2.5%, p=0.005) and fat mass (HP -1.7±3.1%; HC -1.3±2.0%, p=0.006). No differences were seen in any measures in subjects with > mTG. However, a Time x Diet x mTG interaction was observed in changes in hip circumference. Subjects in the HP diet with <mTG experienced a greater reduction in hip circumference (-2.7 ± 4.8%) than those with >mTG levels (-2.4 ± 4.8%, p=0.029) while subjects in the HC diet with >mTG experienced a greater reduction in hip circumference (-3.4 ± 4.8%) than those with <mTG levels (-1.9 ± 3.4%, p=0.029).

Time effects were also observed in systolic and diastolic blood pressures, REE, cholesterol, high density lipoprotein (HDL), low density lipoprotein (LDL) and uric acid. While no time effects were observed with changes in TG, subjects on the HP diet experienced a significantly greater reduction (p=0.048) in TG levels (-5.6 ± 34.0%) than those on the HC (2.0 ± 36.5%) while subjects with >mTG, also experienced a greater reduction (p=0.02) in TG levels (-12.3 ± 29.8%) than those with <mTG (9.1 ± 39.4%).

10.4 CONCLUSION

Results reveal that diet combined with circuit training promotes decreases in waist and hip circumference, weight loss, fat mass and body fat percentage while concomitantly reducing blood pressure, cholesterol and uric acid, and increasing resting energy expenditure. A HP diet promotes greater reductions in weight loss, fat mass and TG levels. Greater reductions in TG levels were experienced by individuals with mTG levels > 125 mg/dL. While a HP diet promotes greater reductions in TG, individuals with TG levels > 125 mg/dL experience greater reductions regardless of diet.

PART V

CELL FUNCTION AND METABOLISM

The Inhibition of the Mammalian DNA Methyltransferase 3a (Dnmt3a) by Dietary Black Tea and Coffee Polyphenols

ARUMUGAM RAJAVELU, ZUMRAD TULYASHEVA, RAKESH JAISWAL, ALBERT JELTSCH, AND NIKOLAI KUHNERT

11.1 BACKGROUND

Black tea is, second only to water, the most consumed beverage globally with an average per capita consumption of around 550 ml per day. The annual production of tea leaves reached a record high in 2008 with a global harvest of 3.75. Mt [1]. Production of dried tea comprises 20% green, 2% oolong and the remainder black. Following black tea, coffee is the third most consumed beverage globally with an annual production of 9.7 Mt and a daily consumption of around 300 ml (data from http://www. fas.usda.gov/, obtained 1st March 2011). Strong epidemiological evidence has repeatedly linked the consumption both black tea [2] and coffee [3,4] to a variety of beneficial health effects, among them is the prevention of multifactorial diseases including cancer, cardiovascular disease and neurological disorders as well as a series of psychoactive responses improving

The Inhibition of the Mammalian DNA Methyltransferase 3a (Dnmt3a) by Dietary Black Tea and Coffee Polyphenols. © *Rajavelu A, Tulyasheva Z, Jaiswal R, Jeltsch A, and Kuhnert N.* BMC Biochemistry *12,16 (2011). doi:10.1186/1471-2091-12-16. Licensed under Creative Commons Attribution 2.0 Generic License, http://creativecommons.org/licenses/by/2.0/.*

alertness, mood and general mental performance [5-8]. Recently, Unilever made an application for a health claim, in which the black tea beverage should supposedly improve mental alertness and focus, based on studies by Nurk et al. with the activities of the two compounds caffeine and L-theanine as the proposed rationale [9]. While epidemiological studies link two causally unrelated events, e. g. a beneficial health effect with the consumption of a certain diet, with a certain statistical probability, the molecular causes of these epidemiological observations are rarely known. In order to rationalize epidemiological observations, a biological target must be identified that is mechanistically linked to the beneficial health effect reported, as well as the specific molecules contained in the diet that interact with the biological target in question at dietary and physiologically relevant concentrations. The search for such matching pairs of biological targets and dietary compound must be considered an exercise of fishing in the dark, however, where enzymes known to be intimately involved in the area in question need to be systematically screened against secondary metabolites known to be produced by the dietary plant in question.

Prompted by reports of Fang and co-workers, who have recently reported the inhibition of DNA methyltransferase 1 (Dnmt1) by a series of dietary polyphenols [10] and work by Lee and co-workers on the inhibition of the same enzyme investigating most notably epi-gallocatechin gallate (EGCG) [11] (the main polyphenolic constituent of green tea) and 5-caffeoyl quinic acid [12] (the main phenolic constituent of the green coffee bean), and Nandakumar, showing the reduction of cellular DNA methylation after admission of (-)-epigallocatechin-3-gallate [13], we decided to screen the interaction of a series of black tea and coffee polyphenols against DNA methyltransferase 3a, another important member of this family of enzymes.

DNA methyltransferases catalyzes methylation of DNA at cytosine residues and play an important role in epigenetic regulation of gene expression, X-chromosome inactivation, genomic imprinting, and development cellular aging and cell differentiation [14,15]. In mammals, DNA methylation is catalyzed mainly by three DNA methyltransferases [15,16]: Dnmt1, Dnmt3a, and Dnmt3b. Dnmt1 has a high preference for hemi-methylated DNA and is essential for maintaining the methylation patterns during each round of DNA replication. On the other hand, Dnmt3a and

Dnmt3b modify both unmethylated and hemimethylated DNA and are responsible for de novo methylation during early development. Errors in DNA methylation contribute to both the initiation and the progression of various cancers [17,18]. In addition, aberrant or missing DNA methylation causes many kinds of diseases which include defects in embryonic development or brain development and neurological defects which are also associated with behavioral changes [19]. Hypermethylation of genes is one of important process in cancer development, typically resulting in the repression of tumor suppressor genes. Preventing the hypermethylation of promoter genes by selective inhibition of methyltransferases could pave a way for cancer treatment [20-22]. Importantly it has been shown that upon use of methyltransferase inhibitors it was possible to reactivate gene silenced by promoter methylation in cancers and thus modulate gene expression. Several efforts are directed at developing small molecules that target DNA methyltransferases and other elements of the machinery, as the proteins that bind to methylated CpG; some are in clinical trials [20-22].

Another important issue of DNA methylation is its function in brain development. Levenson and coworkers showed that Dnmt1 is involved in the formation of hippocampus-dependent long term memory [23]. They found that the promoters for reelin and brain-derived neurotrophic factor (genes implicated in the induction of synaptic plasticity in the adult hippocampus) exhibit rapid and dramatic changes in cytosine methylation when Dnmt1 activity was inhibited. Moreover, DNA methyltransferase inhibitors like 5-aza-2-deoxycytidine blocked the induction of long term potentiation at Schaffer collateral synapses. Furthermore, Dnmt3a-dependent DNA methylation has been reported to influence transcription of neurogenic genes [24]. Additional studies showed that Dnmt1 and Dnmt3a regulate synaptic function in adult forebrain neurons [25] and Dnmt3a affects plasticity of neurons [26].

Changes in the DNA methylation pattern of regions in the hippocampus are associated with behavioral changes in rat [27]. In addition, Dnmt3a has been recently shown to affect the emotional behaviour [26]. Thus, DNA methylation which is already known to be involved in setting up cellular memory is also involved in brain function. The combination of studies in cell lines and in animal models, coupled with data obtained from post-mortem human material provides compelling evidence that aberrant methylation may contribute to psychiatric diseases like schizophrenia and psychosis [28]. Strong epidemiological evidence suggests that particularly for black tea and green tea there is an inverse relation between intake and significant beneficial effects on patients suffering from psychological disorders [2,5-8]. Currently, no accepted rationale on the molecular level exists that can account for these epidemiological findings. Dnmts are a possible biological target for tea dietary polyphenols suggesting a molecular based rationale for the observed biological activities.

11.2 RESULTS

11.2.1 EXPRESSION AND PURIFICATION OF DNMT3A-C

The catalytic domain of Dnmt3a was expressed and purified following an established protocol [29,30]. The purified protein by Ni-NTA affinity chromatography was >90% homogenous as judged from SDS-PAGE stained with colloidal Coomassie Blue (Figure 1).

11.2.2 SELECTION AND PURIFICATION OF BLACK TEA AND COFFEE POLYPHENOLS

Black tea is produced from the young green shoots of the tea plant (Camellia sinensis), which are converted to black tea by fermentation [31]. There are two major processes, the "orthodox" and the "cut-tear-curl." In both, the objective is to achieve efficient disruption of the cellular substructure bringing phenolic compounds present in the green tea leaf, mainly flavan-3-ols otherwise known as catechins, into contact with polyphenol oxidases and activating many other enzymes. The catechin substrates are oxidized and extensively transformed into novel dimeric, oligomeric and polymeric compounds. The chemical composition of black tea brew can be divided into (i) a series of well characterized small molecules including alkaloids (e.g. theobromine and caffeine), carbohydrates and amino acids (including theanine), and a series of glycosylated flavonoids and dimers of catechins including most notably theaflavins together accounting for 30-40% of the dry mass of a typical black tea infusion, and (ii) the heterogeneous and poorly characterized polyphenolic fermentation products accounting for the remaining 60-70% [31]. This material was originally referred to as oxytheotannin and later renamed by Roberts as thearubigins [32].

For this study, we first selected EGCG N1 and (-)-epigallocatechin N4 (from green tea, also on occasions found in black tea at low concentrations) as reference compounds. Next we selected the four most common theaflavin derivatives: theaflavin N2, theaflavin-3-gallate N5, theaflavin 3'-gallate N3 and theaflavin 3, 3'-digallate N6 (see Figure 2) [33,34]. All four compounds are found in black tea infusions at concentrations of around 100 mM, making up 2-3% of the total content of dry mass in typical black tea infusion. Theaflavins are structurally closely related the catechins being formal dimers of EGCG obtained through a two electron oxidation followed by C-C bond formation and a benzylic acid type rearrangement leading to the benztropololone core structure. Next to theaflavins we decided to screen as well two crude thearubigin fractions. We recently proposed that thearubigins contain several thousands polyhydroxylated theaflavin derivatives in equilibrium with their ortho-quinones. Theaflavins were obtained by extraction from black tea infusion followed by purification by preparative HPLC. The purity was assessed by LC-tandem MS. Thearubigins were obtained from black tea infusions using a protocol developed by Roberts [35].

For coffee polyphenols, we selected a range of naturally occurring and synthetic derivatives of chlorogenic acids. Chlorogenic acids (CGAs) are formally hydroxyl-cinnamate esters of quinic acid with a dietary intake of an estimated 2 g per human per day [36]. We recently reported a total of 70 different chlorogenic acids found in green coffee beans and selected some representative examples containing both caffeic acid and ferulic acid substituents [37-40]. Furthermore, we selected a range of epimers of CGAs produced from the original secondary plant metabolites by roasting of the coffee beans. Representative structures are shown in Figure 2. All CGA derivatives were obtained through chemical synthesis unless stated otherwise.

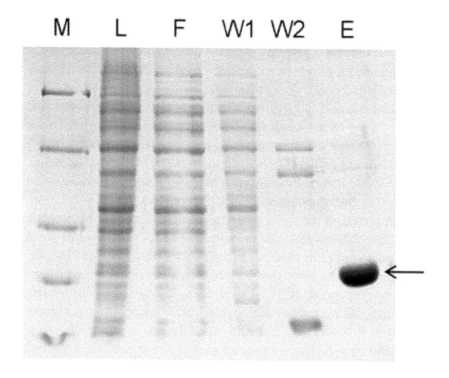

FIGURE 1: Purification of Dnmt3a catalytic domain. Fig. 1 Purified Dnmt3a-C separated on 12% SDS-PAGE gel and stained with colloidal Coomassie Blue. (M, Size marker for 116, 66.2, 45, 35 and 25 kDa; L, crude lysate; F, Flow through; W1, Wash1; W2, Wash2; E, Elution). The purified Dnmt3a-C protein runs at an apparent size of 36 kDa (highlighted with an arrow).

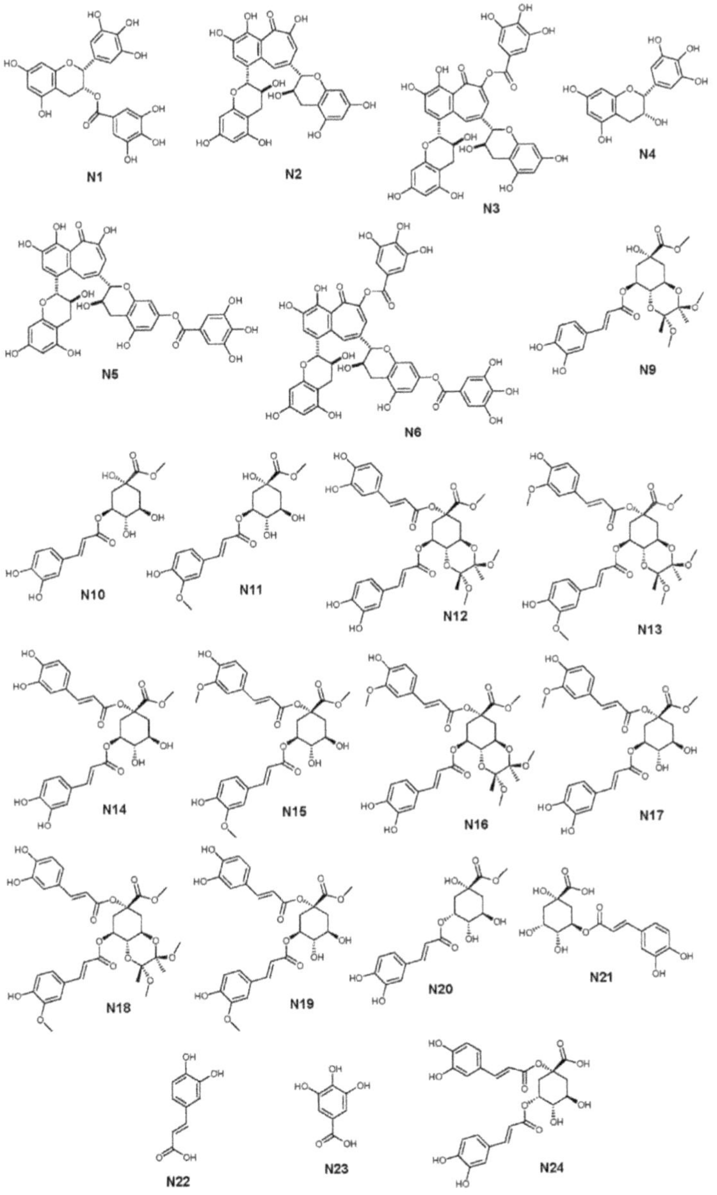

FIGURE 2: Structures of the compounds tested for Dnmt3a-C inhibition.

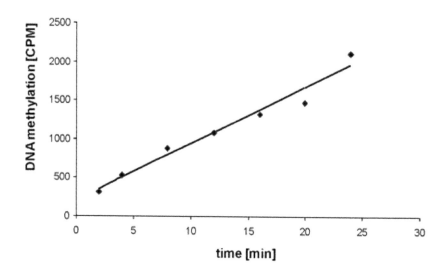

FIGURE 3: Methyltransferase activity of the purified Dnmt3a-C. Example of the methylation kinetics carried out with purified Dnmt3a-C. Initial slopes were determined by linear regression analysis of the initial linear parts of the reaction progress curves.

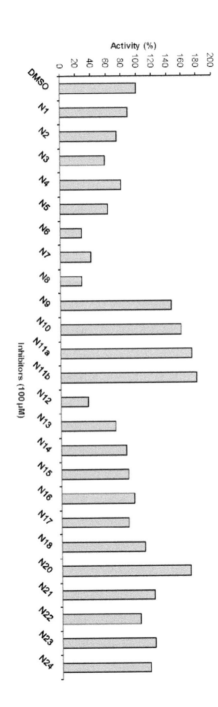

FIGURE 4: Initial screening of the 24 compounds for inhibition of Dnmt3a-C. Dnmt3a-C activity was determined in the presence of 100 µM compound. The control reaction was performed after adding a corresponding volume of DMSO.

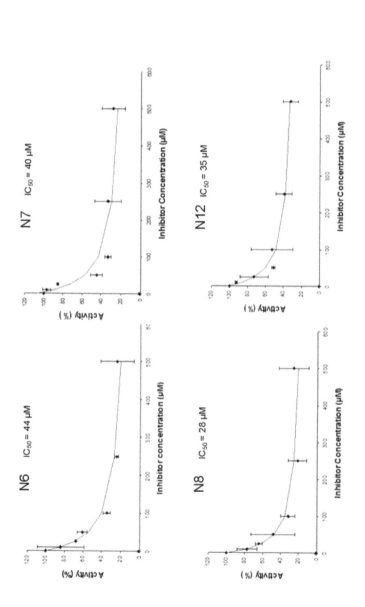

FIGURE 5: Measurement of IC50 values for compounds N6, N7, N8 and N12. For compounds N6, N7, N8 and N11 DNA methylation kinetics were carried out at different concentration of the compounds to determine the IC50 value. The error bars show the maximal deviations in repeated experiments. IC50 values are valid by ±30%.

11.2.3 DNMT3A-C ACTIVITY AND INHIBITORS SCREENING

The purified Dnmt3a-C was catalytically highly active (Figure 3). For an initial screening of the twenty four inhibitor candidates, Dnmt3a-C DNA methylation kinetics were carried out in the presence of 100 µM of compound. Rates of DNA methylation were derived by linear regression of the initial phase of the reaction progress curves. The reaction rates were compared with control reactions carried out after addition of a corresponding volume of DMSO to ensure identical reaction conditions, because DMSO had been shown before to influence the activity of Dnmt3a [41]. As shown in Figure 4, four of the compounds had a substantial inhibitory effect for the in vitro Dnmt3a-C activity (N6-N8 and N12). To determine IC50 values, DNA methylation kinetics were carried out in the presence of variable concentrations of the inhibitors, initial slopes derived and the activity profile analysed by fitting of the experimental data to the equation:

$$A(c_I) = A_0 \times c_I / (IC_{50} + c_I) + BL$$

with: c_I, concentration of the inhibitor; $A(c_I)$, activity in presence of inhibitor at concentration c; A_0, activity in absence of inhibitor; BL, baseline.

As shown in Figure 5, the IC_{50} values for the compounds N6-N8 and N12 were all in the lower µM range.

11.3 DISCUSSION

Lee et al had showed that caffeic acid and chlorogenic acid inhibit the activity of M.SssI and Dnmt1 and decrease the methylation level at the RAR beta promoter gene in the breast cancer cell lines [12]. Furthermore, they have recently described the inhibition of human Dnmt1 by tea flavanoids such as EGCG, catechin and other flavanoids such as quercitin and myristin, observing KI values in the low micromolar range [11]. While Dnmt1 is considered a biological target involved in cancer development its close relative Dnmt3a, investigated in this study, has been linked to both cancer development and mental performance and health. Therefore, any inhibitory interaction between any of the screened dietary polyphenols and Dnmt3a might allow identification of compounds that have a positive effect on cancer prevention and improved mental performance.

11.3.1 BLACK TEA POLYPHENOLS

EGCG (N1) with a reported IC_{50} on Dnmt1 of 0.21 µM and epigallocatechin (N4) showed only weak inhibition of Dnmt3a. A slightly increased activity was observed for theaflavin, theaflavin-3-gallate (N3) and theaflavin 3'-gallate (N5) with the gallated derivatives showing a larger inhibitory effect. Theaflavin 3, 3'-digallate (N6) performed best in this series with a measured IC_{50} value of 44 µM. Similarly, the thearubigin fractions performed well in this test with IC_{50} values of 40 µM and 28 µM, respectively (molarity calculated by assuming an average molecular weight of 800 g/mol). It has to be noted that according to our knowledge this is the first time that a thearubigin fraction (consumed at a level of 1 Mt per annum) has been investigated in an enzyme assay and found to exhibit inhibitory activity. Previous work on thearubigins biological activity had focused on interference with signalling cascades in the anti-inflammatory response [42-45]. Due to the structural similarity of theaflavins and thearubigins (poly-hydroxy theaflavins), the inhibition of Dnmt3a does not come as a complete surprise.

To evaluate any possible biological significance of the IC_{50} values of Dnmt3a inhibition observed here, human pharmacokinetic data need to be consulted. Two published reports address the pharmacokinetic behaviours of theaflavins. Mulder and co-workers report theaflavin concentrations of 4.2 µg l^{-1} in urine 2h after consumption of 1 cup of black tea containing 8.8 mg total theaflavins [46]. Henning reported a concentration of 2 nmol g-1 tissue (if converted around 2 µM) of theaflavin in colon, small intestine, prostate and liver target tissue, with all further three theaflavins N3, N5 and N6 investigated here showing roughly 1 µM, half this value after consumption of one cup of black tea [47]. Although no plasma concentration values are available for theaflavin derivatives, it can be assumed that the plasma concentration is the same order of magnitude if not even higher when compared with concentrations in target tissues. As the average per capita consumption of black tea is around 550 ml or three cups per day, again a higher physiological concentration must be assumed.

From these data it becomes obvious that out of the compounds investigated theaflavin 3, 3'-digallate N6 is a compound showing reasonable bio-availability. These concentration estimate of 2 μM is only roughly by one order of magnitude smaller than the measured IC_{50} values. Assuming consumption of a black tea beverage rich in theaflavins (a maximum of 50 mg l-1 has been determined) or repeated consumption of larger quantities of black tea the measured IC_{50} values for Dnmt3a inhibition, therefore, may have biological significance and inhibition of this enzyme can be expected under physiological conditions after black tea consumption. No data are available on thearubigin pharmacokinetics but since a typical cup of tea contains 60-70% of its dry mass of this mixture of compounds biological significance can as well be assumed.

Two pieces of further work published recently touch on the problem discussed here and are worth highlighting. Firstly, work by Vauzour et al. showed that dietary polyphenols from berries of similar polarity and structure compared to the polyphenols studied here, are able to cross the blood brain barrier [48], therefore suggesting that brain target tissue could be reached by the compounds under investigation. Secondly, recent work by Müller-Harvey et al. reports an accumulation of tea polyphenols in cell nuclei [49], suggesting that not only target tissue but target cell organelles, in which Dnmt3a methylates DNA can indeed be reached by the compounds under investigation.

11.3.2 COFFEE POLYPHENOLS

Out of the twelve chlorogenic acid derivatives screened, seven showed a minor inhibitory effect on Dnmt3a with one compound 1,3-dicaffeoyl-muco-quinic acid diacetal (N12) showing a good IC_{50} value of 35 μM. Since compound N12 is a synthetic derivative, not present in the human diet, this finding has no direct dietary significance. However, the activity of compound N12 clearly indicates that chlorogenic acid derivatives have the potential to inhibit Dnmt3a and this derivative might serve as a lead compound to screen and identify further dietary compound possessing this interesting biological activity.

Interestingly, all compounds showing inhibitory effects are diacyl quinic acids, whereas monoacyl quinic acids showed no effect at all. As a general trend caffeoyl derivatives seem to be more active if compared to feruloyl derivatives and a 1,3-diacyl regiochemistry appears to be favourable. Similarly gallic acid and caffeic acid had no inhibitory effect at all in contrast to the values reported by Lee & Zhu for Dnmt1 inhibition [12]. Despite the structural similarity of these two enzymes a predictive design of inhibitors targeting both classes of enzymes does not seem possible, which can be turned into an advantage considering that the compounds investigated by us and by Lee show remarkable selectivity for either Dnmt1 or Dnmt3a.

11.4 CONCLUSIONS

We have shown that the black tea polyphenols, in particular theaflavin 3, 3'-digallate N6 and thearubigin fraction inhibit Dnmt3a with a physiologically and nutritionally relevant IC50 value and therefore identified a novel biological target that is able to rationalize both anti-carcinogenic activity and mental health and performance related activity of black tea.

11.5 METHODS

11.5.1 EXPRESSION AND PURIFICATION OF DNMT3A-C

The mouse Dnmt3a C-terminal domain was expressed and purified as described [29,30]. The purity of protein was determined on 12% SDS-PAGE gel stained with colloidal Coomassie Blue (Figure 1). Protein concentration was determined from the absorbance at 280 nm using an extinction coefficient of 39290 M^{-1} cm^{-1}.

11.5.2 DNA METHYLTRANSFERASE ACTIVITY ASSAY

Kinetics of Dnmt3a-C was analyzed by using a Biotin-Avidin methylation kinetics assay basically as described [50] using a biotinylated oligonucleotide substrate and [methyl-3H]AdoMet.

FP3   5'-TTGCACTCTCCTCCCGGAAGTCCCAGCTTC-3'   FP3-Bt 5'-Bt-GAAGCTGGGACTTCCGGGAGGAGAGTGCAA-3'; The oligonucleotides were annealed by heating to 86°C for some minutes and slowly cooling down to ambient temperature. The methylation reactions were carried out in methylation buffer [20 mM HEPES pH 7.2, 1 mM EDTA, 50 mM KCl, 25 mg/ml bovine serum albumin (BSA)] at 37°C, using 1 µM substrate DNA, 0.76 µM AdoMet and 2.5 µM Dnmt3a-C. After the methylation reaction, the oligonucleotides were immobilized at various time points on an avidin-coated microplate. The incorporation of [3H] into the DNA was quenched by addition of an excess of unlabeled AdoMet to the binding buffer. Subsequently, unreacted AdoMet was removed by washing five times with PBST containing 0.5 M NaCl. The immobilized DNA was digested with a non-specific endonuclease to release the radioactivity from the microplate. After digestion, 120 µl of the reaction mixture were transferred to a fresh microplate and 160 µl of Microscint-PS scintillation fluid (Perkin Elmer) was added to each well. Finally, the amount of methyl groups transferred to the DNA and the solution obtained after nucleolytic digestion was quantified by using the TopCount NXT liquid scintillation counter. To determine the initial slope, the data were fitted by linear regression of the initial part of the reaction progress curves.

All the inhibitors were prepared in the DMSO at 5 mM stock. For the screening purpose 100 µM concentrations of the inhibitors were used in the reaction mixture. To detemine the apparent IC50 value for the potential inhibitors, different concentration of the inhibitors were used in the reaction mixture (10 µM, 25 µM, 50 µM, 100 µM, 250 µM, 500 µM). The different concentrations of the inhibitors were incubated with Dnmt3a protein for 10 min at room temp. The reaction was started by adding substrate and cofactor and further incubated at 37°C for another 10 min then the reaction was stopped by adding excess unlabelled AdoMet. The DMSO was used as control in each experimental setup to exclude the possible inhibition effect from the DMSO itself. All the inhibitor kinetics was done at duplicate and standard error was calculated for the two experimental values.

11.5.3 ISOLATION AND SYNTHESIS OF INHIBITORS

EGCG N1 and (-)-epigallocatechin (N4), theaflavin (N2), theaflavin-3-gallate (N3), theaflavin 3'-gallate (N5) and theaflavin 3, 3'-digallate (N6) were from black tea obtained using published procedures (Figure 2) [33,51]. Thearubigin fractions (N7 and N8) were obtained from black tea and characterised using published procedures [33,51]. All chlorogenic acid derivatives were obtained by synthesis using published procedures (Figure 2) [52].

REFERENCES

1. Poulter S: [http://dailymail.co.uk] Daily Mail Online. 2008.
2. Gardner EJ, Ruxton CHS, Leeds AR: Black tea - helpful or harmful? A review of the evidence. European Journal of Clinical Nutrition 2007, 61(1):3-18.
3. Higdon JV, Frei B: Coffee and health: A review of recent human research. Critical Reviews in Food Science and Nutrition 2006, 46(2):101-123.
4. Dorea JG, da Costa THM: Is coffee a functional food? British Journal of Nutrition 2005, 93(6):773-782.
5. Hamer M, Williams ED, Vuononvirta R, Gibson EL, Steptoe A: Association between coffee consumption and markers of inflammation and cardiovascular function during mental stress. Journal of Hypertension 2006, 24(11):2191-2197.
6. Franco R: Coffee and mental health. Atencion Primaria 2009, 41(10):578-581.
7. Hozawa A, Kuriyama S, Nakaya N, Ohmori-Matsuda K, Kakizaki M, Sone T, Nagai M, Sugawara Y, Nitta A, Tomata Y, et al.: Green tea consumption is associated with lower psychological distress in a general population: the Ohsaki Cohort 2006 Study. American Journal of Clinical Nutrition 2009, 90(5):1390-1396.
8. de Mejia EG, Ramirez-Mares MV, Puangpraphant S: Bioactive components of tea: Cancer, inflammation and behavior. Brain Behavior and Immunity 2009, 23(6):721-731.
9. Nurk E, Refsum H, Drevon CA, Tell GS, Nygaard HA, Engedal K, Smith AD: Intake of Flavonoid-Rich Wine, Tea, and Chocolate by Elderly Men and Women Is Associated with Better Cognitive Test Performance. Journal of Nutrition 2009, 139(1):120-127.

10. Fang MZ, Chen DP, Yang CS: Dietary polyphenols may affect DNA methylation. Journal of Nutrition 2007, 137(1):223S-228S.

11. Lee WJ, Shim JY, Zhu BT: Mechanisms for the inhibition of DNA methyltransferases by tea catechins and bioflavonoids. Molecular Pharmacology 2005, 68(4):1018-1030.

12. Lee WJ, Zhu BT: Inhibition of DNA methylation by caffeic acid and chlorogenic acid, two common catechol-containing coffee polyphenols. Carcinogenesis 2006, 27(2):269-277.

13. Nandakumar V, Vaid M, Katiyar SK: (-)-Epigallocatechin-3-gallate reactivates silenced tumor suppressor genes, Cip1/p21 and p16INK4a, by reducing DNA methylation and increasing histones acetylation in human skin cancer cells. Carcinogenesis 2011.

14. Klose RJ, Bird AP: Genomic DNA methylation: the mark and its mediators. Trends Biochem Sci 2006, 31(2):89-97.

15. Jurkowska RZ, Jurkowski TP, Jeltsch A: Structure and function of mammalian DNA methyltransferases. Chembiochem 2011, 12(2):206-222.

16. Jeltsch A: Beyond Watson and Crick: DNA methylation and molecular enzymology of DNA methyltransferases. Chembiochem 2002, 3(4):274-293.

17. Jones PA, Baylin SB: The epigenomics of cancer. Cell 2007, 128(4):683-692.

18. Feinberg AP, Tycko B: The history of cancer epigenetics. Nat Rev Cancer 2004, 4(2):143-153.

19. Egger G, Liang GN, Aparicio A, Jones PA: Epigenetics in human disease and prospects for epigenetic therapy. Nature 2004, 429(6990):457-463.

20. Yoo CB, Jones PA: Epigenetic therapy of cancer: past, present and future. Nature Reviews Drug Discovery 2006, 5(1):37-50.

21. Gal-Yam EN, Saito Y, Egger G, Jones PA: Cancer epigenetics: modifications, screening, and therapy. Annu Rev Med 2008, 59:267-280.

22. Kelly TK, De Carvalho DD, Jones PA: Epigenetic modifications as therapeutic targets. Nat Biotechnol 2010, 28(10):1069-1078.

23. Levenson JM, Sweatt JD: Epigenetic mechanisms: a common theme in vertebrate and invertebrate memory formation. Cellular and Molecular Life Sciences 2006, 63(9):1009-1016.

24. Wu H, Coskun V, Tao J, Xie W, Ge W, Yoshikawa K, Li E, Zhang Y, Sun YE: Dnmt3a-dependent nonpromoter DNA methylation facilitates transcription of neurogenic genes. Science 2010, 329(5990):444-448.

25. Feng J, Zhou Y, Campbell SL, Le T, Li E, Sweatt JD, Silva AJ, Fan G: Dnmt1 and Dnmt3a maintain DNA methylation and regulate synaptic function in adult forebrain neurons. Nat Neurosci 2010, 13(4):423-430.

26. LaPlant Q, Vialou V, Covington HE, Dumitriu D, Feng J, Warren BL, Maze I, Dietz DM, Watts EL, Iniguez SD, et al.: Dnmt3a regulates emotional behavior and spine plasticity in the nucleus accumbens. Nat Neurosci 2010, 13(9):1137-1143.

27. Weaver ICG, Cervoni N, Champagne FA, D'Alessio AC, Sharma S, Seckl JR, Dymov S, Szyf M, Meaney MJ: Epigenetic programming by maternal behavior. Nature Neuroscience 2004, 7(8):847-854.

28. Feng J, Fan G: The role of DNA methylation in the central nervous system and neuropsychiatric disorders. Int Rev Neurobiol 2009, 89:67-84.

29. Gowher H, Jeltsch A: Molecular enzymology of the catalytic domains of the Dnmt3a and Dnmt3b DNA methyltransferases. J Biol Chem 2002, 277(23):20409-20414.

30. Jurkowska RZ, Anspach N, Urbanke C, Jia D, Reinhardt R, Nellen W, Cheng X, Jeltsch A: Formation of nucleoprotein filaments by mammalian DNA methyltransferase Dnmt3a in complex with regulator Dnmt3L. Nucleic Acids Res 2008, 36(21):6656-6663.

31. Drynan JW, Clifford MN, Obuchowicz J, Kuhnert N: The chemistry of low molecular weight black tea polyphenols. Natural Product Reports 2010, 27(3):417-462.

32. Roberts EAH, Catrwright RA, Oldschool M: J Sci Food Agric. 1959, 8:72-80.

33. Kuhnert N, Drynan JW, Obuchowicz J, Clifford MN, Witt M: Mass spectrometric characterization of black tea thearubigins leading to an oxidative cascade hypothesis for thearubigin formation. Rapid Communications in Mass Spectrometry 2010, 24(23):3387-3404.

34. Kuhnert N: Unraveling the structure of the black tea thearubigins. Archives of Biochemistry and Biophysics 2010, 501(1):37-51.

35. Roberts EAH, Myers M: J Sci Food Agric. 1959, 10:167-179.

36. Clifford MN: Chlorogenic acids and other cinnamates - nature, occurrence and dietary burden. Journal of the Science of Food and Agriculture 1999, 79(3):362-372.

37. Clifford MN, Johnston KL, Knight S, Kuhnert N: Hierarchical scheme for LC-MSn identification of chlorogenic acids. Journal of Agricultural and Food Chemistry 2003, 51(10):2900-2911.

38. Clifford MN, Knight S, Surucu B, Kuhnert N: Characterization by LC-MSn of four new classes of chlorogenic acids in green coffee beans: Dimethoxycinnamoylquinic acids, diferuloylquinic acids, caffeoyl-dimethoxycinnamoylquinic acids, and feruloyl-dimethoxycinnamoylquinic acids. Journal of Agricultural and Food Chemistry 2006, 54(6):1957-1969.

39. Jaiswal R, Patras MA, Eravuchira PJ, Kuhnert N: Profile and Characterization of the Chlorogenic Acids in Green Robusta Coffee Beans by LC-MSn: Identification of Seven New Classes of Compounds. Journal of Agricultural and Food Chemistry 2010, 58(15):8722-8737.

40. Jaiswal R, Sovdat T, Vivan F, Kuhnert N: Profiling and Characterization by LC-MSn of the Chlorogenic Acids and Hydroxycinnamoylshikimate Esters in Mate (Ilex paraguariensis). Journal of Agricultural and Food Chemistry 2010, 58(9):5471-5484.

41. Yokochi T, Robertson KD: Dimethyl sulfoxide stimulates the catalytic activity of de novo DNA methyltransferase 3a (Dnmt3a) in vitro. Bioorg Chem 2004, 32(4):234-243.

42. Lin YL, Tsai SH, Lin-Shiau SY, Ho CT, Lin JK: Theaflavin-3,3'-digallate from black tea blocks the nitric oxide synthase by down-regulating the activation of NF-kappa B in macrophages. European Journal of Pharmacology 1999, 367(2-3):379-388.

43. Bhattacharya U, Halder B, Mukhopadhyay S, Giri AK: Role of oxidation-triggered activation of JNK and p38 MAPK in black tea polyphenols induced apoptotic death of A375 cells. Cancer Science 2009, 100(10):1971-1978.

44. Jochmann N, Lorenz M, von Krosigk A, Martus P, Bohm V, Baumann G, Stangl K, Stangl V: The efficacy of black tea in ameliorating endothelial function is equivalent to that of green tea. British Journal of Nutrition 2008, 99(4):863-868.

45. Lorenz M, Urban J, Engelhardt U, Baumann G, Stangl K, Stangl V: Green and black tea are equally potent stimuli of NO production and vasodilation: new insights into tea ingredients involved. Basic Research in Cardiology 2009, 104(1):100-110.

46. Vermeer MA, Mulder TP, Molhuizen HO: Theaflavins from black tea, especially theaflavin-3-gallate, reduce the incorporation of cholesterol into mixed micelles. J Agric Food Chem 2008, 56(24):12031-12036.

47. Henning SM, Aronson W, Niu YT, Conde F, Lee NH, Seeram NP, Lee RP, Lu JX, Harris DM, Moro A, et al.: Tea polyphenols and theaflavins are present in prostate tissue of humans and mice after green and black tea consumption. Journal of Nutrition 2006, 136(7):1839-1843.

48. Vauzour D, Rendeiro C, Corona G, Williams C, Spencer JEP: Polyphenol commun. 2010, 1:70-71.

49. Müller-Harvey I, Botchway S, Feucht W, Polster J, Burgos P, Parker A: Planta Medica. 2010, 76:1167.

50. Roth M, Jeltsch A: Biotin-avidin microplate assay for the quantitative analysis of enzymatic methylation of DNA by DNA methyltransferases. Biological Chemistry 2000, 381(3):269-272.

51. Kuhnert N, Clifford MN, Muller A: Oxidative cascade reactions yielding polyhydroxy-theaflavins and theacitrins in the formation of black tea thearubigins: Evidence by tandem LC-MS. Food & Function 2010, 1(2):180-199.

52. Kuhnert N, Jaiswal R, Eravuchira P, El-Abassy RM, von der Kammer B, Materny A: Scope and limitations of principal component analysis of high resolution LC-TOF-MS data: the analysis of the chlorogenic acid fraction in green coffee beans as a case study. Analytical Methods 2011, 3(1):144-155.

CHAPTER 12

Treatment of Human Muscle Cells with Popular Dietary Supplements Increase Mitochondrial Function and Metabolic Rate

ROGER A. VAUGHAN, RANDI GARCIA-SMITH,
MIGUEL A. BARBERENA, MARCO BISOFFI, KRISTINA TRUJILLO,
AND CAROLE A. CONN

12.1 BACKGROUND

Obesity is an increasingly prevalent morbidity with nearly two thirds of adult Americans overweight, half of whom are obese [1]. Obesity-related health issues have been reported to increase healthcare costs by an estimated $147 billion annually [2]. Over the past decade, chemical and behavioral interventions that favorably modify metabolic rate have been central to obesity research.

Several over-the-counter dietary supplements claim to increase metabolic rate and enhance fatty acid catabolism. Of the available over-the-

Treatment of Human Muscle Cells with Popular Dietary Supplements Increase Mitochondrial Function and Metabolic Rate. © Vaughan RA, Garcia-Smith R, Barberena MA, Bisoffi M, Trujillo K, and Conn CA. Current Status and Prospects of Biodiesel Production from Microalgae. Nutrition & Metabolism **9**,101 (2012), doi:10.1186/1743-7075-9-101. Licensed under Creative Commons Attribution 2.0 Generic License, http://creativecommons.org/licenses/by/2.0/.

counter (OTC) dietary supplements, OxyElite Pro (OEP) produced by USP Laboratory and Cellucore Super HD (CHD) are purported to increase metabolic rate and fat metabolism [3].

OEP has been shown to increase markers of fat mobilization, metabolic rate (measured via indirect calorimetry), and reduce bodyweight and body fat (estimated via Dual-energy X-ray absorptiometry) in healthy young subjects following ingestion [3,4]. A key ingredient 1,3-dimethylamylamine (also known as germanium, geranamine or DMAA) has been implicated for potential contraindications. DMAA is purported to increase both systolic and diastolic blood pressure in young and healthy men and women immediately following ingestion, although these observations have been inconsistent during longer treatments with OEP in young men [4-7]. DMAA is also purported to cause false-positive results for amphetamines on select immunoassays, a profound implication for athletes with sanctioned governing bodies [8]. CHD, a newer dietary supplement has very limited research regarding safety or efficacy. CHD is advertised to increase metabolic rate and decrease fatty acid synthesis. Many of the ingredients including caffeine have been previously linked to increased metabolic rate. Moreover, because supplements commonly contain a variety of ingredients in proprietary blend forms, and few controlled studies have been performed to address the metabolic effects at the cellular level, further work is needed to identify possible metabolic effects. This work specifically addresses the effects that treatment with OEP or CHD supplements have on metabolism in human skeletal muscle cells.

Peroxisome proliferator-activated receptor coactivator 1 alpha (PGC-1α) is a transcriptional coactivator that is essential for mitochondrial biosynthesis and activates genes that regulate energy homeostasis and metabolism [9-11]. PGC-1α increases fatty acid oxidation through increased peroxisome proliferator-activated receptor alpha (PPARα) expression, which increases forkhead box protein 1 (FOXO1), nuclear respiratory factors 1 and 2 (NRF1/2) and other factors influencing fat oxidation [12-14]. PGC-1α is also an important signaling molecule in the activation and regulation of gluconeogenesis, which is likely mediated through FOXO1 and estrogen-related receptor alpha (ERR-α) [12,15-18]. Thus, PGC-1α modifies metabolic rate and expression of genes involved in gluconeogenesis, fat oxidation and mitochondrial biosynthesis [12-18].

Clinically, the relationship between low PGC-1α expression and type II diabetes/obesity has been identified [19-22]. Low PGC-1α is also associated with reduced expression of oxidative phosphorylation genes, decreasing fatty acid oxidation and energy utilization [19,23,24]. Treatment with the PGC-1α stimulator Rosiglitazone (through binding and activating PPARγ) increased mitochondrial density and function, while improving insulin sensitivity [25]. Further evidence suggests that an increase in PGC-1α (independent of Rosiglitazone) can improve insulin sensitivity and improve muscle function [25]. It has also been identified that PGC-1α is essential for the recovery from the diminished ATP caused by chemical uncoupling as evidenced by the lack of recovery in PGC-1α null cells and animals [26].

Treatment with potent research-grade chemicals such as 2,4-dinitrophenol (DNP) and p-trifluromethoxy phenylhydrazone (FCCP) have been shown to induce PGC-1α in fibroblasts [26]. Moreover, our laboratory recently identified that treatment with DNP or caffeine can induce PGC-1α, and increase both metabolic rate and mitochondrial content in muscle cells suggesting that commercially available metabolic stimulators might have similar effects [27]. The well documented effects of PGC-1α on metabolism suggest that modulation of PGC-1α expression is a potential strategy for altering metabolic rate.

Purpose. This work seeks to explore effects of treatment with OTC dietary supplements on mitochondrial and glycolytic metabolism in skeletal muscle cells. Human rhabdomyosarcoma cells are a naturally immortalized cell model, frequently used for making inferences about muscle tissue adaptations [27-31]. We show that treatment of muscle cells with OEP or CHD at varied doses induce PGC-1α mRNA and protein in a dose and time sensitive manner. We also illustrate that treatment with either OEP or CHD increase mitochondrial content. This work identifies for the first time the effects that several OTC diet supplements have on mitochondria content and cell metabolism in muscle cells.

12.2 METHODS

12.2.1 CELL CULTURE

Homo sapiens rhabdomyosarcoma cells were purchased from ATCC (Manassas, VA). Cells were cultured in Dulbecco's Modified Eagle's Medium (DMEM) containing 4500mg/L glucose and supplemented with 10% heat-inactivated fetal bovine serum (FBS) and 100U/mL penicillin/streptomycin, in a humidified 5% CO_2 atmosphere at 37°C. Trypsin-EDTA at 0.25% was used to detach the cells for splitting and re-culturing. Stock Oxy Elite ProTM (OEP) from USP Labs (Dallas, TX) and stock Cellucore HDTM (CHD) from Cellucore (Bryan, TX) purchased over the counter were diluted to 2 dilutions that contain equivalent ingredient by weight; high dose containing 90 µg/ml or low dose containing 45 µg/ml. Dose and exposure times were determined through pilot experiments to significantly increase PGC-1α (data not shown). Final concentration of ethanol was 0.1% for all treatments.

12.2.2 QUANTITATIVE REAL-TIME POLYMERASE CHAIN REACTIONS (QRT-PCR)

Cells were seeded in 6-well plates at a density of 1×10^6 cells/well, treated and incubated as described above for 12 or 24 hours. Following incubation, total RNA was extracted using RNeasy Kit from Qiagen (Valencia, CA) and total RNA was quantified by Nanodrop spectrophotometry. RNA (5000 ng/sample) was denatured at 75°C for 3 minutes and cDNA was synthesized using random decamers and Moloney murine leukemia virus reverse transcriptase (MMLVRT) from the Retroscript™ RT kit from Ambion (Austin, TX) for 60 minutes at 42°C and the enzyme denatured at 92°C for 10 minutes. PCR primers were designed using Primer Express software from Invitrogen (Carlsbad, CA) and synthesized by Integrated DNA Technologies (IDT; Coralville, IA). For PGC-1α, the forward primer was 5'-ACCAAACCCACAGAGAACAG-3' and the reverse primer was 5'-GGGTCAGAGGAAGAGATAAAGTTG-3'. Amplification of PGC-1α was normalized to the housekeeping gene, TATA Binding Protein (TBP). For TBP, the forward primer was 5'-CACGAACCACGGCACTGATT-3' and the reverse primer was 5'-TTTTCTTGCTGCCAGTCTGGAC-3'. qRT-PCR reactions were performed in triplicate using the LightCycler 480 real-time PCR system from Roche Applied Science, (Indianapolis, IN). SYBR Green based PCR was performed in triplicate using an estimated 800 ng of cDNA per well to ensure a strong signal; final primer concentrations were 10 μM in a total volume of 30μl. The following cycling parameters were used: 95°C for 10 minutes followed by 45 cycles of 95°C for 15 seconds, and 60°C for one minute. Relative expression levels were determined by the change in crossing points of reaction amplification (ΔΔCp method) between PGC-1α and TBP for each treatment compared with the control group [32].

12.2.3 FLOW CYTOMETRY

Cells were plated in 6-well plates at a density of 1.2×10^6 cells/well treated in triplicate and incubated as previously described above for 24 hours. Following treatment, the media was removed and the cells were re-suspended in pre-warmed media with 200 nM Mitotracker Green from Life Technologies (Carlsbad, CA) and incubated for 45 minutes in a humidified 5% CO_2 atmosphere at 37°C. The cells were pelleted, the media with Mitotracker was removed and the cells were suspended in pre-warmed media. Group mean fluorescence was measured using Facscalibur filtering 488nm.

12.2.4 MICROSCOPY AND IMMUNOHISTOCHEMISTRY

Chamber-slides from BD Bioscience (Sparks, MD), were seeded with 5000 cells/well. To verify PGC-1α protein expression, cells were cultured and treated for 24 hours as described above. Cells were fixed using 3.7% formaldehyde in media, permiabilized with PBS with 0.1% Triton 100X from Sigma (St. Louis, MO) for 10 minutes and blocked for 1 hour with PBS with 0.1% Triton 100X and 3.0% BSA from Sigma (St. Louis, MO). Cells were stained with an anti-PGC-1α primary polyclonal antibody from

Santa Cruz Biotechnologies (Santa Cruz, CA) at 1:200 dilution in PBS with 0.1% BSA overnight. The cells were rinsed with PBS with 0.1% Triton 100X and 3.0% BSA, and secondary anti-rabbit AlexFluor 633 antibody from Invitrogen (Carlsbad, CA) was applied in 1:200 dilution. Slides were mounted with Prolong Gold with DAPI from Invitrogen (Carlsbad, CA) and cured overnight. Cells were imaged using the Axiovert 25 microscope with AxioCam MRc from Zeiss (Thornwood, NY). To verify increased mitochondrial content, the cells were then stained with Mitotracker 200 nM from Invitrogen (Carlsbad, CA) for 45 minutes, and fixed in 3.7% formaldehyde in pre-warmed media. Cells were mounted, cured and imaged as described above.

12.2.5 METABOLIC ASSAY

Cells were seeded overnight in 24-well culture plate from SeaHorse Bioscience (Billerica, MA) at density 5 x 10^5 cells/well, treated and incubated for 24 hours as described above. Following treatment, culture media was removed and replaced with XF Assay Media from SeaHorse Bioscience (Billerica, MA) containing 4500mg/L glucose free of CO_2 and incubated at 37°C. Per manufactures' protocol, SeaHorse injection ports were loaded with oligomycin, an inhibitor of ATP synthase which induces maximal glycolytic metabolism and reveals endogenous proton leak (mitochondrial uncoupling) at a final concentration 1.0 µM. Oligomycin addition was followed by the addition of carbonyl cyanide p-[trifluoromethoxy]-phenyl-hydrazone (FCCP), an uncoupler of electron transport that induces peak oxygen consumption (an indirect indicator of peak oxidative metabolism) at final concentration 1.25 µM. Rotenone was then added in 1.0 µM final concentration to reveal non-mitochondrial respiration and end the metabolic reactions [33,34]. Extracellular acidification, an indirect measure of glycolytic capacity, and oxygen consumption, a measure of oxidative metabolism was measured using the SeaHorse XF24 Extracellular Analyzer from SeaHorse Bioscience (Billerica, MA). SeaHorse XF24 Extracellular Analyzer was run using 8 minute cyclic protocol commands (mix for 3 minutes, let stand 2 minutes, and measure for 3 minutes) in triplicate.

12.2.6 WST1 METABOLIC ASSAY

WST-1 is a widely used reagent that is metabolized by mitochondrial de-hydrogenases through consumption of reduction potential resulting in the fluorescence indicating changes in metabolism, cell proliferation and vi-ability. We used the WST-1 assay as an indirect measure of cellular reduc-tion potential which indicates cellular energy status. Cells were seeded in 96-well plates at density 5,000 cells/well and grown over night. Cells were then treated with either ethanol control, or one of the designated OTC supplement treatments and incubated as previously described above for 24 hours. Media and treatment were removed at each time point and media containing 10% WST1 reagent was added to each well and incubated as previously described above. Fluorescence was measured 1 hour follow-ing WST1 addition using Wallac Victor3V 1420 Multilabel Counter from PerkinElmer (Waltham, MA).

12.2.7 CELL VIABILITY

Cells were seeded for 24 hours in 6-well plate with density 1.2×10^6 cells/well and treated in triplicate and incubated as previously described above for 24 hours. Trypan blue from Sigma (St Louis, MO) exclusion staining was used to assess cell number and viability using Countess from Invitro-gen (Carlsbad, CA) cell quantification system.

12.2.8 STATISTICAL ANALYSIS

PGC-1α expression, flow cytometry, metabolic assays and cell viability were analyzed using ANOVA with Dunnett's post hoc test and pairwise comparisons were used to compare treatments with control. WST1 cell metabolism analysis was performed using ANOVA and pairwise compari-sons comparing treatments with control following reciprocal transforma-tion of group log fluorescence and data normalization to control (control metabolic rate = 1). Values of $p < 0.05$ indicated statistical significance in all tests and Prism from GraphPad (La Jolla, CA) was used to perform all statistical analyses.

12.3 RESULTS AND DISCUSSION

12.3.1 PGC-1α EXPRESSION AND MITOCHONDRIAL CONTENT

PGC-1α RNA was significantly induced following treatment for 12 or 24 hours with OEPHigh and Low or CHDLow compared with the control group (Figure 1A and B, respectively). PGC-1α protein was imaged using confocal microscopy and quantified with ImageJ using 7 cells per treatment (n = 7) which revealed significantly elevated PGC-1α protein following treatment for 24 hours with OEPHigh or CHDHigh (Figure 1C and D). Flow cytometry with Mitotracker staining showed that treatment for 24 hours with either OEP or CHD significantly increased mitochondrial content compared with ethanol control (Figure 2A). We used confocal microscopy to verify flow cytometry observations that OEP or CHD treated cells had increased mitochondrial content (Figure 1B and C).

12.3.2 GLYCOLYTIC METABOLISM

In order to quantify changes in glycolytic metabolism, we measured extracellular acidification rate (ECAR), which was significantly increased in cells treated with high dose supplements (Figure 3A). Treatment with either OEPHigh or CHDHigh significantly increased basal glycolysis compared with control (Figure 3B). Treatment with either OEPHigh or CHDHigh also significantly increased peak glycolytic capacity compared with control (Figure 3C). OEPHigh or CHDHigh treatment also significantly elevated glycolytic use during induction of peak oxidative metabolism compared with the control group (Figure 3D). Cells treated with low-dose OEP or CHD did not have greater basal or peak glycolytic capacity compared with control (data not shown).

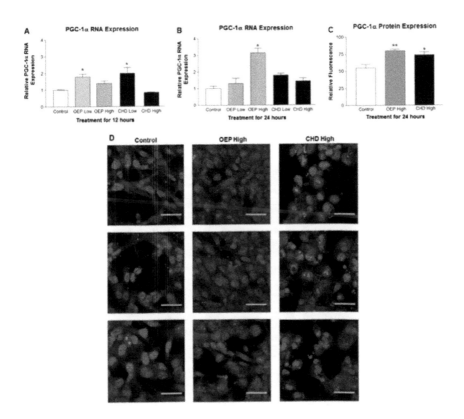

FIGURE 1: Group mean PGC-1α RNA and protein expression ±SD.A- Relative induction of PGC-1α RNA following treatment of rhabdomyosarcoma cells with OEP or CHD at 90 µg/ml (High) or 45 µg/ml (Low) for 12 hours compared with ethanol control (control =1). B- Relative induction of PGC-1α RNA following similar treatment for 24 hours. C-Immunofluorescent quantification of PGC-1α protein expression from confocal microscopy in (D) using 7 cells per treatment (n = 7) following treatment with OEP or CHD at 90 µg/ml (High) for 24 hours compared with ethanol control. D- Confocal microscopy of PGC-1α protein expression of cells treated as described above for 24 hours. NOTES: Dapi nuclear stain (blue), PGC-1α protein (red), and green line represents 50 µm. * indicates p < 0.05, ** indicates p < 0.01 and *** indicates p < 0.001.

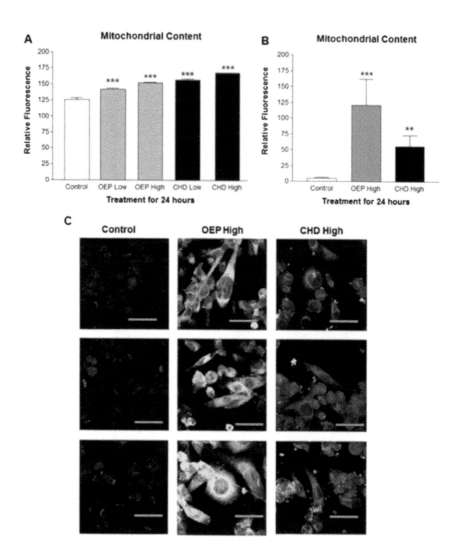

FIGURE 2: Group mean mitochondrial content ±SD. A- Flow cytometric group mean log fluorescence of rhabdomyosarcoma cells measured in triplicate using 10,000 cells per sample, treated with OEP or CHD at 90 μg/ml (High) or 45 μg/ml (Low). B- Immunofluorescent quantification of cells treated as described above for 24 hours using 7 cells per treatment (n = 7). C- Confocal microscopy images of cells treated as described above for 24 hours. NOTES: Dapi nuclear stain (blue), mitochondrial stain (green), and red line represents 50 μm. * indicates $p < 0.05$, ** indicates $p < 0.01$ and *** indicates $p < 0.001$.

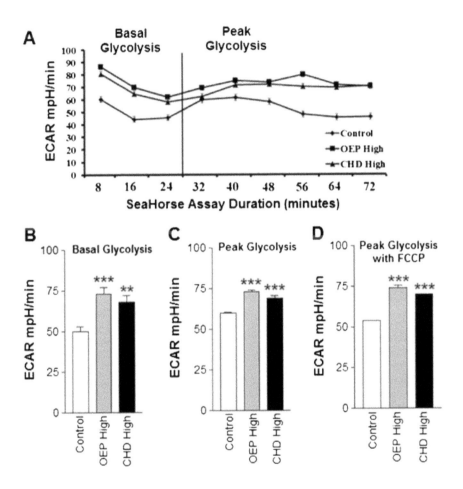

FIGURE 3: Group mean glycolytic metabolism ±SD.A- Extracellular acidification rate (ECAR) of rhabdomyosarcoma cells treated with either ethanol control (final concentration = 0.1%) or OEP or CHD at 90 µg/ml (High) for 24 hours. B- Baseline ECAR following treatment described above. C- Peak ECAR following addition of oligomycin, an inhibitor of oxidative phosphorylation. D- Peak ECAR during peak oxygen consumption rate (OCR) following addition of carbonyl cyanide p-[trifluoromethoxy]-phenyl-hydrazone (FCCP), a mitochondrial uncoupling agent, in addition to previously added oligomycin. NOTES: * indicates p < 0.05, ** indicates p < 0.01, and *** indicates p < 0.001 compared with control.

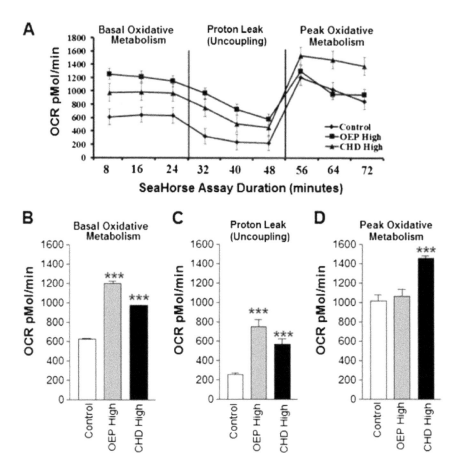

FIGURE 4: Group mean oxidative metabolism ±SD. A- Oxygen consumption rate (OCR) of rhabdomyosarcoma cells treated with either ethanol control (final concentration = 0.1%) or OEP or CHD at 90 μg/ml (High). B- Baseline OCR following treatment described above. C- OCR during peak ECAR following addition of oligomycin, an inhibitor of oxidative phosphorylation. D- Peak OCR following addition of carbonyl cyanide p-[trifluoromethoxy]-phenyl-hydrazone (FCCP), a mitochondrial uncoupling agent, in addition to previously added oligomycin. NOTES: * indicates $p < 0.05$, ** indicates $p < 0.01$, and *** indicates $p < 0.001$ compared with control.

12.3.3 OXIDATIVE METABOLISM

In order to quantify changes in oxidative metabolism, we measured oxygen consumption rate (OCR) which was also elevated in cells treated with high dose OEP or CHD (Figure 4A). Treatment with either OEPHigh or CHDHigh significantly increased basal oxidative metabolism compared with control (Figure 4B). Treatment with either OEPHigh or CHDHigh also significantly increased oxidative metabolism during peak glycolytic capacity compared with control, an indicator of increased endogenous mitochondrial uncoupling (or potentially chemically-induced uncoupling) (Figure 4C). Unexpectedly, only treatment with CHDHigh significantly elevated peak oxidative metabolism compared with the control group (Figure 4D). Cells treated with low-dose OEP or CHD did not consistently exhibit greater basal or peak oxidative metabolism capacity compared with control (data not shown). Only OEPHigh exhibited significantly greater oxygen consumption compared with the control following addition of rotenone, an indirect measure of non-mitochondrial respiration (data not shown).

12.3.4 CELLULAR OXIDATIVE RELIANCE

To quantify changes in cellular reliance on oxidative metabolism, we compared ECAR versus OCR. Oxidative reliance, indicated by a ratio of OCR:ECAR was significantly increased in cells treated with high dose OEP or CHD (Figure 5A). Specifically, cells treated with OEPHigh or CHDHigh demonstrated significantly greater reliance on oxidative metabolism during basal measurements and during peak glycolysis compared with the control group (Figure 5C and D, respectively).

12.3.5 BASAL METABOLIC RATE

To quantify changes in metabolic rate, we compared ECAR versus OCR. Cells treated with OEPHigh or CHDHigh exhibited a significantly greater basal ECAR and OCR, indicating increased metabolic rate (Figure 6A).

Moreover, cellular metabolism indicated by WST-1 assay was significantly increased in cells treated with either dietary supplement compared with control. Both OEP and CHD increased metabolism in a dose-dependent manner compared with ethanol control (Figure 6B). To assess the effect of supplement treatment on cell viability, we used Trypan blue exclusion measured by Countess Cell Counter. Following 24 hour of treatment with either ethanol or either of the supplements at either dose, cell viability was not statistically different from control (data not shown).

This work identified several effects that dietary supplements OEP and CHD have on metabolism and cellular adaptation in skeletal muscle cells. First, we illustrated that both dietary supplements increase mitochondrial density and content in part through induction of metabolic transcription factor PGC-1α. Flow cytometry and microscopy experiments verified that cells treated with either supplement exhibited significantly greater mitochondrial content versus ethanol control. We also employed metabolic experiments which identified that both oxidative and total metabolism were significantly increased by both supplements compared with control. WST-1 assay observations in combination with insignificant changes in cell viability also support the observation that both treatments significantly increased cellular metabolism.

Collectively, these results support the notion that stimulators of PGC-1α and mitochondrial metabolism may have implications for many different metabolic diseases such as diabetes and obesity [19-21,25]. Previously, our lab has demonstrated that polyunsaturated fatty acids are dietary constituents which stimulate PGC-1α and mitochondrial metabolism in skeletal muscle cells [35]. Our current observations demonstrate the stimulatory effect of dietary supplements and support several of the previous examinations of the effects of OEP on whole body metabolism [3,4]. This report is among the first to demonstrate that the relatively new dietary supplement CHD also increases metabolism. Lastly, this report is among the first to show an increase in mitochondrial content following treatment with OEP and CHD.

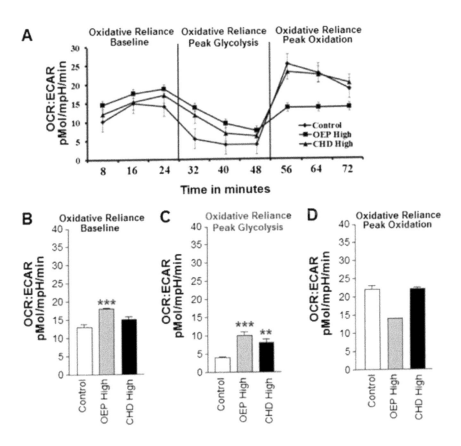

FIGURE 5: Group mean oxidative reliance ±SD. A- Metabolic reliance expressed as the ratio of oxygen consumption rate (OCR) extracellular acidification rate (ECAR) OCR:ECAR illustrating metabolic rate. B Oxidative reliance represented by a ratio of OCR:ECAR at baseline. C- Oxidative reliance during peak glycolysis. D- Oxidative reliance during peak oxidation. NOTES: * indicates $p < 0.05$, ** indicates $p < 0.01$, and *** indicates $p < 0.001$ compared with control.

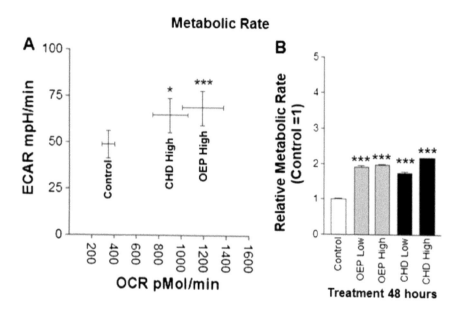

FIGURE 6: A- Group mean basal metabolic rate ± SD of rhabdomyosarcoma cells treated with either ethanol control (final concentration = 0.1%) or OEP or CHD at 90 µg/ml (High) plotted as ECAR versus OCR (data from SeaHorse assay).B- Change in metabolic rate derived from reciprocal log fluorescence from WST1 metabolic assay with control = 1 indicating change in reduction potential within each well treated with either 0.1% ethanol or OEP or CHD at 90 µg/ml (High) or 45 µg/ml (Low). NOTES: * indicates $p < 0.05$, ** indicates $p < 0.01$, and *** indicates $p < 0.001$ compared with control.

12.3.7 LIMITATIONS

Our experiments were performed using a cancerous myoblast model known to favor glycolytic metabolism as opposed to oxidative metabolism, which may make our observations less generalizable to healthy mature skeletal muscle. Additionally, we prepared culture media per ATCC recommendations containing high glucose; an environment previously shown to induce insulin resistance in L6-myotubes [36]. It would be interesting to examine the effects these supplements elicit from mature, differentiated myotubes in various media. Further research is needed to elucidate the depth of cellular effects that these potent metabolic stimulators may have on metabolic diseases such as diabetes.

12.4 CONCLUSION

Manipulation of mitochondrial function provides many appealing possibilities for the treatment of several diseases including obesity. This work illustrates that, although much of the biochemical framework has been completed, there is still much to learn about the role that PGC-1α can play in the treatment of various disease states. Further research is needed to identify both behavior modifications and chemical agents that can elicit an increase in mitochondria, because of the role that diminished mitochondria play in chronic diseases. It appears that increased PGC-1α expression and activation accompanied by increased mitochondrial biosynthesis, may be a viable treatment option for diseases such as obesity.

REFERENCES

1. Flegal KM, Carroll MD, Ogden CL, Johnson CL: Prevalence and trends in obesity among US adults, 1999–2000. JAMA 2002, 288:1723-1727.
2. Study estimates medical cost of obesity May Be as high as $147 billion annually. http://www.cdc.gov/media/pressrel/2009/r090727.htm.
3. McCarthy CG, Farney TM, Canale RE, Alleman RJ Jr, Bloomer RJ: A finished dietary supplement stimulates lipolysis and metabolic rate in young Men and women. Nutrition and Metabolic Insights 2011, 5:23.

4. McCarthy CG, Canale RE, Bloomer RJ, Alleman RJ Jr: Biochemical and anthropometric effects of a weight loss dietary supplement in healthy Men and women. Nutr Metab Insights 2011, 5:13.

5. Bloomer RJ, Harvey IC, Farney TM, Bell ZW, Canale RE: Effects of 1,3-dimethylamylamine and caffeine alone or in combination on heart rate and blood pressure in healthy Men and women. Phys Sportsmed 2011, 39:111-120.

6. Whitehead PN, Schilling BK, Farney TM, Bloomer RJ: Impact of a dietary supplement containing 1,3-dimethylamylamine on blood pressure and bloodborne markers of health: a 10-week intervention study. Nutr Metab Insights 2012, 5:33.

7. Farney TM, McCarthy CG, Canale RE, Alleman RJ Jr, Bloomer RJ: Hemodynamic and hematologic profile of healthy adults ingesting dietary supplements containing 1,3-dimethylamylamine and caffeine. Nutr Metab Insights 2011, 5:1.

8. Vorce SP, Holler JM, Cawrse BM, Magluilo J Jr: Dimethylamylamine: a drug causing positive immunoassay results for amphetamines. J Anal Toxicol 2011, 35:183-187.

9. Wu H, Kanatous SB, Thurmond FA, Gallardo T, Isotani E, Bassel-Duby R, Williams RS: Regulation of mitochondrial biogenesis in skeletal muscle by CaMK. Science 2002, 296:349-352.

10. Esterbauer H, Oberkofler H, Krempler F, Patsch W: Human peroxisome proliferator activated receptor gamma coactivator 1 (PPARGC1) gene: cDNA sequence, genomic organization, chromosomal localization, and tissue expression. Genomics 1999, 62:98-102.

11. Knutti D, Kaul A, Kralli A: A tissue-specific coactivator of steroid receptors, identified in a functional genetic screen. Mol Cell Biol 2000, 20:2411-2422.

12. Yoon JC, Puigserver P, Chen GX, Donovan J, Wu ZD, Rhee J, Adelmant G, Stafford J, Kahn CR, Granner DK, et al.: Control of hepatic gluconeogenesis through the transcriptional coactivator PGC-1. Nature 2001, 413:131-138.

13. Puigserver P, Rhee J, Donovan J, Walkey CJ, Yoon JC, Oriente F, Kitamura Y, Altomonte J, Dong HJ, Accili D, Spiegelman BM: Insulin-regulated hepatic gluconeogenesis through FOXO1-PGC-1 alpha interaction. Nature 2003, 423:550-555.

14. Vega RB, Huss JM, Kelly DP: The coactivator PGC-1 cooperates with peroxisome proliferator-activated receptor alpha in transcriptional control of nuclear genes encoding mitochondrial fatty acid oxidation enzymes. Mol Cell Biol 2000, 20:1868-1876.

15. Herzig S, Long FX, Jhala US, Hedrick S, Quinn R, Bauer A, Rudolph D, Schutz G, Yoon C, Puigserver P, et al.: CREB regulates hepatic gluconeogenesis through the coactivator PGC-1 (vol 413, pg 179, 2001). Nature 2001, 413:652-652.

16. Rhee J, Ge HF, Yang WL, Fan M, Handschin C, Cooper M, Lin JD, Li C, Spiegelman BM: Partnership of PGC-1 alpha and HNF4 alpha in the regulation of lipoprotein metabolism. J Biol Chem 2006, 281:14683-14690.

17. Alaynick WA: Nuclear receptors, mitochondria and lipid metabolism. Mitochondrion 2008, 8:329-337.

18. Shao D, Liu Y, Liu X, Zhu L, Cui Y, Cui A, Qiao A, Kong X, Liu Y, Chen Q, et al.: PGC-1 beta-Regulated mitochondrial biogenesis and function in myotubes is mediated by NRF-1 and ERR alpha. Mitochondrion 2010, 10:516-527.

19. Patti ME, Butte AJ, Crunkhorn S, Cusi K, Berria R, Kashyap S, Miyazaki Y, Kohane I, Costello M, Saccone R, et al.: Coordinated reduction of genes of oxidative metabolism in humans with insulin resistance and diabetes: Potential role of PGC1 and NRF1. Proc Natl Acad Sci USA 2003, 100:8466-8471.

20. Semple RK, Crowley VC, Sewter CP, Laudes M, Christodoulides C, Considine RV, Vidal-Puig A, O'Rahilly S: Expression of the thermogenic nuclear hormone receptor coactivator PGC-1 alpha is reduced in the adipose tissue of morbidly obese subjects. Int J Obes 2004, 28:176-179.

21. Yang XL, Enerback S, Smith U: Reduced expression of FOXC2 and brown adipogenic genes in human subjects with insulin resistance. Obes Res 2003, 11:1182-1191.

22. Chowdhury SKR, Dobrowsky RT, Femyhough P: Nutrient excess and altered mitochondrial proteome and function contribute to neurodegeneration in diabetes. Mitochondrion 2011, 11:845-854.

23. Mootha VK, Handschin C, Arlow D, Xie XH, Sihag S, Yang WL, Altshuler D, Puigserver P, Patterson N, St Pierre J, et al.: Err alpha and Gabpa/b specify PGC-1 alpha-dependent oxidative phosphorylation gene expression that is altered in diabetic muscle (vol 101, pg 6570, 2004). Proc Natl Acad Sci USA 2005, 102:10405-10405.

24. Schreiber SN, Emter R, Hock MB, Knutti D, Cardenas J, Podvinec M, Oakeley EJ, Kralli A: The estrogen-related receptor alpha (ERR alpha) functions in PPAR gamma coactivator 1 alpha (PGC-1 alpha)-induced mitochondrial biogenesis. Proc Natl Acad Sci USA 2004, 101:6472-6477.

25. Benton CR, Bonen A: PGC-1alpha expression correlates with skeletal muscle fibre type and fuel handling proteins. FASEB J 2005, 19:A1121-A1121.

26. Rohas LM, St-Pierre J, Uldry M, Jager S, Handschin C, Spiegelman BM: A fundamental system of cellular energy homeostasis regulated by PGC-1 alpha. Proc Natl Acad Sci USA 2007, 104:7933-7938.

27. Vaughan RA, Garcia-Smith R, Bisoffi M, Trujillo KA, Conn CA: Effects of caffeine on metabolism and mitochondria biogenesis in rhabdomyosarcoma cells compared with 2,4-dinitrophenol. Nutr Metab Insights 2012, 5:59.

28. Carrasco MA, Muller M, Cardenas C, Quest A, Behrens MI, Jaimovich E: Role of calcium and PKC in CREB phosphorylation induced by depolarization in skeletal muscle cells in culture. Soc Neurosci Abstr 2001, 27:1296.

29. Armoni M, Quon MJ, Maor G, Avigad S, Shapiro DN, Harel C, Esposito D, Goshen Y, Yaniv I, Karnieli E: PAX3/Forkhead homolog in rhabdomyosarcoma oncoprotein activates glucose transporter 4 gene expression in vivo and in vitro. J Clin Endocrinol Metabol 2002, 87:5312-5324.

30. Araki M, Maeda M, Motojima K: Hydrophobic statins induce autophagy and cell death in human rhabdomyosarcoma cells by depleting geranylgeranyl diphosphate. Eur J Pharmacol 2012, 674:95-103.

31. Singh P, Kohr D, Kaps M, Blaes F: Skeletal muscle MHC I expression: implications for statin-induced myopathy. Muscle Nerve 2010, 41:179-184.

32. Pfaffl MW: A new mathematical model for relative quantification in real-time RT-PCR. Nucleic Acids Res 2001, 29:e45.

33. Wikstrom JD, Sereda SB, Stiles L, Elorza A, Allister EM, Neilson A, Ferrick DA, Wheeler MB, Shirihai OS: A novel high-throughput assay for islet respiration reveals uncoupling of rodent and human islets. PLoS One 2012, 7:e33023.
34. Giulivi C, Ross-Inta C, Horton AA, Luckhart S: Metabolic pathways in anopheles stephensi mitochondria. Biochem J 2008, 415:309-316.
35. Vaughan RA, Garcia-Smith R, Bisoffi M, Conn CA, Trujillo KA: Conjugated linoleic acid or omega 3 fatty acids increase mitochondrial biosynthesis and metabolism in skeletal muscle cells. Lipids Health Dis 2012, 11:142.
36. Huang C, Somwar R, Patel N, Niu WY, Torok D, Klip A: Sustained exposure of L6 myotubes to high glucose and insulin decreases insulin-stimulated GLUT4 translocation but upregulates GLUT4 activity. Diabetes 2002, 51:2090-2098.

Author Notes

CHAPTER 1

Acknowledgments
The Bellagio Report was approved by the Participants of the Bellagio Meeting on "Healthy Agriculture, Healthy Nutrition, Healthy People": Conference Chairs: Artemis P. Simopoulos, Peter G. Bourne, Ole Faergeman; Honorary Co-Chairs: Dan Glickman, Philip R. Lee; Participants: Jon Andrus, Gail C. Christopher, Sarath Gopalan, Russell Jaffe, Richard J. Johnson, Leonidas Koskos, Philippe Legrand, Duo Li, Ascensión Marcos, Anne M. Peniston, Sam Rehnborg, Laura A. Schmidt, Ian Scott, Kraisid Tontisirin, Dan L. Waitzberg.

CHAPTER 2

Author Contribution
A. C. Fahed and G. M. Nemer designed the narrative review and the different sections of the manuscript, supervised the work, and compiled the final draft. A. M. El-Hage-Sleiman and T. I. Farhat performed the literature search, designed the figures, and wrote equally different sections of the manuscript. A. K. M. El-Hage-Sleiman and T. I. Farhat contributed equally to this paper. All authors read and approved the final manuscript.

Acknowledgments
The authors would like to thank all the Nemer Lab team members for their support during the preparation of this work. They would also like to thank Dr. Amy Zenger from the Writing Center at the American University of Beirut for editing the manuscript.

CHAPTER 3

Acknowledgments

We thank James Graham and Marinelle Nunez for excellent technical support and the nursing staff at the University of California, Davis, Clinical Research Center for their dedicated nursing support. We also acknowledge and thank Janet Peerson for expert advice on the statistical analysis and Dr. Richard J. Havel for his expert advice and editing. This work was supported by National Institutes of Health (NIH)/National Heart, Lung, and Blood Institute Grant 1R01 HL09133. The project also received support from Grant UL1 RR024146 from the National Center for Research Resources, a component of the NIH, andNIH Roadmap for Medical Research.

Disclosure Summary

K.L.S., A.A.B., V.M., G.C., T.H.F., V.L., R.I.M., N.L.K, and P.J.H. have nothing to disclose. K.N. has consulted for Denka Seiken Co., Tokyo, Japan, and Otsuka Pharmaceutical Co., Ltd., Tokyo, Japan. T.N. was previously employed by Otsuka Pharmaceutical Co., Ltd., Tokyo, Japan. Y.I. is employed by Denka Seiken Co., Tokyo, Japan.

CHAPTER 4

Competing Interests

The authors declare that they have no competing interests.

Author Contributions

FD has participated as PhD student and main investigator. BL, DC and GL contributed to the conception of the protocol, data analysis and manuscript drafting. LR and VS measured all biologic data. RC had responsibilities in daily diet. ED managed physical activity. GL has revised and given final approval of the manuscript. All authors have read and approved the final manuscript.

Acknowledgements

This work was supported by PRES Université Blaise Pascal – Clermont II – Laboratory AME2P Metabolic Adaptations to Exercise in Physiological and Pathological conditions and by the thermal baths of Chatel-Guyon.

The study was funded by the Heart and Diseases Foundation (Fondation Coeur et Artères) 59200 Loos, France; http://www.fondacoeur.com.
Our thanks to Jeffrey Watts for help with manuscript English proof reading.

CHAPTER 5

Competing Interests
The authors declare that they have no competing interests.

Author Contributions
GP gave classes, nutritional orientations and antropometric measures in all sample, writing of the article and critically reviewed the article. RS participated of protocol desing, did cognitive measures and reviewed the manuscript. YJ and NN carried out of the statistical analysis. IO was responsible for DA diagnostics. CQ and PH was involved in the protocol and the study desing, analysis and writing of the article. All authors read and approved the final manuscript.

Acknowledgements
We want to acknowledge: CAPES - Ministry of Education and Abbott Laboratories for their contribution to the realization of this work and the opportunity given to provide all supplementation that was distributed to all patients freely.

CHAPTER 6

Competing Interests
The authors declare that they have no competing interests.

Author Contributions
All authors were involved in the design of the study; DR and AM conducted the research; RR coordinated the laboratory analyses; DR performed the statistical analysis, wrote the manuscript, and had primary responsibility for the final content. All authors read and approved the final manuscript.

Acknowledgements

We appreciate the efforts of the following individuals: staff at ICDDR,B and Shimantik who supported and implemented the study, including Taufiq Rahman, Sultana Mahabbat-e Khoda, Evana Akhtar, Eliza Roy, Ashish Chowdhury, and Kazi Moksedur Rahman; Reinhold Vieth (Mount Sinai Hospital, Toronto) for performing the measurement of 25-hydroxyvitamin D concentrations and verifying the concentration of the vitamin D3 supplement; Brendon Pezzack for assistance with manuscript preparation. Thank you to Diasorin Inc. (Stillwater, MN) for donating the kits used in the Liaison Total assay, and to Popular Pharmaceuticals Inc. (Dhaka, Bangladesh) for supplying the Vigantol Oil.

Research was supported by funding from the Center for Global Health, Johns Hopkins University (A. Baqui) and the Department of International Health at The Johns Hopkins Bloomberg School of Public Health (R. Black). D. Roth was supported by training awards from The Alberta Heritage Foundation for Medical Research and The Canadian Institutes of Health Research. Sponsors did not have any role in study design, data collection, data analysis, data interpretation, writing of the manuscript, or the decision to submit the manuscript for publication.

CHAPTER 7

Funding

The authors thank the British Heart Foundation (grant PG/09/023) and the UK Medical Research Council (MRC; grant G0601653) for funding this work. ADH is a British Heart Foundation Senior Research Fellow (Award FS05/125). EH is a Department of Health (UK) Public Health Career Scientist. This work was undertaken at the Centre for Paediatric Epidemiology and Biostatistics, which benefits from funding support from the MRC in its capacity as the MRC Centre of Epidemiology for Child Health. Research at the University College London Institute of Child Health and Great Ormond Street Hospital for Children NHS Trust benefits from R&D funding received from the NHS Executive. No funding bodies had any role in study design, data collection and analysis, decision to publish, or preparation of the manuscript.

Competing Interests

LTH is currently supported by a Canada Institute of Research (CIHR) Fellowship award. CC has received honoraria and consulting fees from Amgen, Eli Lilly, Medtronic, Merck, Novartis, and Servier. WM is an employee of synlab laboratory services GmbH. Synlab offers vitamin D testing. TJW is on the scientific advisory board for Diasorin Inc. and has received research support from them. JCW is 90% employed by GlaxoSmithKline (GSK) whilst maintaining a 10% appointment at London School of Hygiene & Tropical Medicine (LSHTM), and holds GSK shares. All other authors declare that no competing interests exist.

Acknowledgments

We thank Alana Cavadino (UCL ICH) for technical assistance in formatting the figures. Members of the Genetic Investigation of Anthropometric Traits (GIANT) Consortium include: Elizabeth K Speliotes, Cristen J Willer, Sonja I Berndt, Keri L Monda, Gudmar Thorleifsson, Anne U Jackson, Hana Lango Allen, Cecilia M Lindgren, Jian'an Luan, Reedik Mägi, Joshua C Randall, Sailaja Vedantam, Thomas W Winkler, Lu Qi, Tsegaselassie Workalemahu, Iris M Heid, Valgerdur Steinthorsdottir, Heather M Stringham, Michael N Weedon, Eleanor Wheeler, Andrew R Wood, Teresa Ferreira, Robert J Weyant, Ayellet V Segrè, Karol Estrada, Liming Liang, James Nemesh, Ju-Hyun Park, Stefan Gustafsson, Tuomas O Kilpeläinen, Jian Yang, Nabila Bouatia-Naji, Tõnu Esko, Mary F Feitosa, Zoltán Kutalik, Massimo Mangino, Soumya Raychaudhuri, Andre Scherag, Albert Vernon Smith, Ryan Welch, Jing Hua Zhao, Katja K Aben, Devin M Absher, Najaf Amin, Anna L Dixon, Eva Fisher, Nicole L Glazer, Michael E Goddard, Nancy L Heard-Costa, Volker Hoesel, Jouke-Jan Hottenga, Åsa Johansson, Toby Johnson, Shamika Ketkar, Claudia Lamina, Shengxu Li, Miriam F Moffatt, Richard H Myers, Narisu Narisu, John R B Perry, Marjolein J Peters, Michael Preuss, Samuli Ripatti, Fernando Rivadeneira, Camilla Sandholt, Laura J Scott, Nicholas J Timpson, Jonathan P Tyrer, Sophie van Wingerden, Richard M Watanabe, Charles C White, Fredrik Wiklund, Christina Barlassina, Daniel I Chasman, Matthew N Cooper, John-Olov Jansson, Robert W Lawrence, Niina Pellikka, Inga Prokopenko, Jianxin Shi, Elisabeth Thiering, Helene Alavere, Maria T S Alibrandi, Peter Almgren, Alice M Arnold, Thor Aspelund, Larry D Atwood, Beverley Balkau, Anthony J Balmforth, Amanda J Bennett, Yoav Ben-Shlomo, Richard N Bergman, Sven Bergmann, Heike Biebermann, Alexandra I F Blakemore, Tanja Boes, Lori L Bonnycastle, Stefan R Bornstein, Morris J Brown, Thomas A Buchanan, Fabio Busonero, Harry Campbell, Frances-co P Cappuccio, Christine Cavalcanti-Proença, Yii-Der Ida Chen, Chih-Mei Chen, Peter S Chines, Robert Clarke, Lachlan Coin, John Connell, Ian N M Day, Martin den Heijer, Jubao Duan, Shah Ebrahim, Paul Elliott, Roberto Elosua, Gudny Eiriksdottir, Michael R Erdos, Johan G Eriksson, Maurizio F Facheris, Stephan B Felix, Pamela Fischer-Posovszky, Aaron R Folsom, Nele Friedrich, Nelson B Freimer, Mao Fu, Stefan Gaget, Pablo V Gejman, Eco J C Geus, Christian Gieger, Anette P Gjesing, Anuj Goel, Philippe Goyette, Harald Grallert, Jürgen Gräßler, Danielle M Green-awalt, Christopher J Groves, Vilmundur Gudnason, Candace Guiducci, Anna-Liisa Hartikainen, Neelam Hassanali, Alistair S Hall, Aki S Havu-linna, Caroline Hayward, Andrew C Heath, Christian Hengstenberg, An-drew A Hicks, Anke Hinney, Albert Hofman, Georg Homuth, Jennie Hui, Wilmar Igl, Carlos Iribarren, Bo Isomaa, Kevin B Jacobs, Ivonne Jarick, Elizabeth Jewell, Ulrich John, Torben Jørgensen, Pekka Jousilahti, Antti Jula, Marika Kaakinen, Eero Kajantie, Lee M Kaplan, Sekar Kathiresan, Johannes Kettunen, Leena Kinnunen, Joshua W Knowles, Ivana Kolcic, Inke R König, Seppo Koskinen, Peter Kovacs, Johanna Kuusisto, Peter Kraft, Kirsti Kvaløy, Jaana Laitinen, Olivier Lantieri, Chiara Lanzani, Lenore J Launer, Cecile Lecoeur, Terho Lehtimäki, Guillaume Lettre, Ji-anjun Liu, Marja-Liisa Lokki, Mattias Lorentzon, Robert N Luben, Bar-bara Ludwig, Paolo Manunta, Diana Marek, Michel Marre, Nicholas G Martin, Wendy L McArdle, Anne McCarthy, Barbara McKnight, Thom-as Meitinger, Olle Melander, David Meyre, Kristian Midthjell, Grant W Montgomery, Mario A Morken, Andrew P Morris, Rosanda Mulic, Ju-lius S Ngwa, Mari Nelis, Matt J Neville, Dale R Nyholt, Christopher J O'Donnell, Stephen O'Rahilly, Ken K Ong, Ben Oostra, Guillaume Paré, Alex N Parker, Markus Perola, Irene Pichler, Kirsi H Pietiläinen, Carl G P Platou, Ozren Polasek, Anneli Pouta, Suzanne Rafelt, Olli Raitakari, Nigel W Rayner, Martin Ridderstråle, Winfried Rief, Aimo Ruokonen, Neil R Robertson, Peter Rzehak, Veikko Salomaa, Alan R Sanders, Manjinder S Sandhu, Serena Sanna, Jouko Saramies, Markku J Savolainen, Susann Scherag, Sabine Schipf, Stefan Schreiber, Heribert Schunkert, Kaisa Silander, Juha Sinisalo, David S Siscovick, Jan H Smit, Nicole Soranzo, Ulla Sovio, Jonathan Stephens, Ida Surakka, Amy J Swift, Mari-Liis Tammesoo, Jean-Claude Tardif, Maris Teder-Laving, Tanya M Teslovich, John R Thompson, Brian Thomson, Anke Tönjes, Tiinamaija Tuomi, Joyce B J van Meurs, Gert-Jan van Ommen, Vincent Vatin, Jorma Viikari, Sophie Visvikis-Siest, Veronique Vitart, Carla I G Vogel, Benjamin F Voight, Lindsay L Waite, Henri Wallaschofski, G Bragi Walters, Elisabeth Widen, Susanna Wiegand, Sarah H Wild, Gonneke Willemsen, Daniel R Witte, Jacqueline C Witteman, Jianfeng Xu, Qunyuan Zhang, Lina Zgaga, Andreas Ziegler, Paavo Zitting, John P Beilby, I Sadaf Farooqi, Johannes Hebebrand, Heikki V Huikuri, Alan L James, Mika Kähönen, Douglas F Levinson, Fabio Macciardi, Markku S Nieminen, Claes Ohlsson, Lyle J Palmer, Paul M Ridker, Michael Stumvoll, Jacques S Beckmann, Heiner Boeing, Eric Boerwinkle, Dorret I Boomsma, Mark J Caulfield, Stephen J Chanock, Francis S Collins, L Adrienne Cupples, George Davey Smith, Jeanette Erdmann, Philippe Froguel, Henrik Grönberg, Ulf Gyllensten, Per Hall, Torben Hansen, Tamara B Harris, Andrew T Hattersley, Richard B Hayes, Joachim Heinrich, Frank B Hu, Kristian Hveem, Thomas Illig, Marjo-Riitta Jarvelin, Jaakko Kaprio, Fredrik Karpe, Kay-Tee Khaw, Lambertus A Kiemeney, Heiko Krude, Markku Laakso, Debbie A Lawlor, Andres Metspalu, Patricia B Munroe, Willem H Ouwehand, Oluf Pedersen, Brenda W Penninx, Annette Peters, Peter P Pramstaller, Thomas Quertermous, Thomas Reinehr, Aila Rissanen, Igor Rudan, Nilesh J Samani, Peter E H Schwarz, Alan R Shuldiner, Timothy D Spector, Jaakko Tuomilehto, Manuela Uda, André Uitterlinden, Timo T Valle, Martin Wabitsch, Gérard Waeber, Nicholas J Wareham, Hugh Watkins, James F Wilson, Alan F Wright, M Carola Zillikens, Nilanjan Chatterjee, Steven A McCarroll, Shaun Purcell, Eric E Schadt, Peter M Visscher, Themistocles L Assimes, Ingrid B Borecki, Panos Deloukas, Caroline S Fox, Leif C Groop, Talin Haritunians, David J Hunter, Robert C Kaplan, Karen L Mohlke, Jeffrey R O'Connell, Leena Peltonen, David Schlessinger, David P Strachan, Cornelia M van Duijn, H-Erich Wichmann, Timothy M Frayling, Unnur Thorsteinsdottir, Gonçalo R Abecasis, Inês Barroso, Michael Boehnke, Kari Stefansson, Kari E North, Mark I McCarthy, Joel N Hirschhorn, Erik Ingelsson and Ruth J F Loos.

Author Contributions

Conceived and designed the experiments: KSV DJB MM OR JV BDM EAM NJW TM TL WHO CC TJW MRJ JCW ADH EH. Performed the experiments: KSV DJB CL ET SP LTH JDC KHL ZD RL ARW KM LZ LMY JD MCK KJ VS BDM EAS DKH TL PK CC RJFL JCW ADH EH. Analyzed the data: KSV DJB CL ET SP LTH JDC KHL ZD RL ARW KM LZ LMY JD MCK KJ VS BDM EAS DKH TL PK CC RJFL JCW ADH EH. Contributed reagents/materials/analysis tools: KSV DJB SP LTH KM LMY MM JD MK NA OR JV SBK HM EI LB LL VS BDM KH EAS DKH TL PK CC WM RJFL CP MRJ JCW ADH EH. Wrote the first draft of the manuscript: KSV DJB EH. Contributed to the writing of the manuscript: KSV DJB EH. ICMJE criteria for authorship read and met: KSV DJB CL ET SP LTH JDC ZD RL DKH ARW KM LV LZ LMY MM JD MK MEK KJ NA OR JV KKL LF HM EI LB LL ML VS HC MD BDM KHH AP ALH EAS ET AJ NJW CO TMF SBK TDS JBR TL WHO PK CC WM CP RJFL TJW MRJ JCW ADH EH. Agree with manuscript results and conclusions: KSV DJB CL ET SP LTH JDC ZD RL DKH ARW KM LV LZ LMY MM JD MK MEK KJ NA OR JV KKL LF HM EI LB LL ML VS HC MD BDM KHH AP ALH EAS ET AJ NJW CO TMF SBK TDS JBR TL WHO PK CC WM CP RJFL TJW MRJ JCW ADH EH. Obtained the data: KSV DJB SP LTH ZD RL KM LV MM MK MCK KJ NA OR JV SBK LF HM EI LB LL ML VS HC MD BDM AP AH ET NJW CO TMF DKH JBR TL WHO PK CC WM CP ADH MRJ EH. Provided the administrative, technical, or material support: KSV DJB KM MK OR JV SBK LF HM EI LL VS TMF DKH TL CP EH. Supervised the study: JD OR JV SBK HC MD BDM EAS AJ ET TS TMF JBR TL TJW MRJ JCW ADH EH.

CHAPTER 8

Conflicts of Interest

The authors declare no conflict of interest.

CHAPTER 9

Conflict of Interest

The authors declare no conflict of interest.

CHAPTER 10

Acknowledgement

We would like to thank Jean Jitomir, Monica Serra, Jen Moreillon, Erika Deike, Geoffrey Hudson, and Mike Greenwood who assisted in data collection on the first cohort of subjects that participated in this study when the ESNL was located at Baylor University. This study was supported by Curves International, Waco, TX.

CHAPTER 11

Author Contributions

AR, ZT and RJ conducted and analyzed the experiments. NK and AJ participated in the design of the study and in data analysis and interpretation and drafted the manuscript. All authors read and approved the final manuscript.

Acknowledgements

The authors thank Jacobs University Bremen for a scholarship to Rakesh Jaiswal. Technical assistance from Ms. Anja Müller is gratefully acknowledged. This work was supported by DFG grant JE 252/6.

CHAPTER 12

Competing Interests

Roger Vaughan: Was previously employed by General Nutrition Center, a retail agency that sells tested supplements. All authors and contributors declare no other competing interests.

Author Contribution

RV: Performed and over saw all experiments, primary author of manuscript, producer of experimental design, and received Research, Project and Travel Grant to support this project. RG: Assisted with laboratory procedures. MAB: Assisted with laboratory procedures. MB: Assisted with experimental design and manuscript production. KT: Financially supported experiment, assisted with experimental design and manuscript production. CAC: Assisted with experimental design and manuscript production. All authors read and approved the final manuscript.

Acknowledgements

Funding: Funding was provided by the University of New Mexico Summer 2012 Office of Graduate Studies Research, Project and Travel Grant, and through Department of Biochemistry and Molecular Biology Faculty Research Allocation Funds provided by Kristina Trujillo Ph.D. We would like to thank the University of New Mexico Department of Biochemistry and Molecular Biology for their assistance in this work.

Index